Information Circular 9513

Proceedings of the Second American Conference on Human Vibration

Chicago, IL
June 4–6, 2008

Technical Editors: Farid Amirouche, Ph.D., and Alan G. Mayton, C.M.S.P., P.E.

Document Editor: Robert J. Tuchman

DEPARTMENT OF HEALTH AND HUMAN SERVICES
Centers for Disease Control and Prevention
National Institute for Occupational Safety and Health
Pittsburgh Research Laboratory
Pittsburgh, PA

June 2009

This document is in the public domain and may be freely copied or reprinted.

Disclaimer

Mention of any company or product does not constitute endorsement by the National Institute for Occupational Safety and Health (NIOSH). In addition, citations to Web sites external to NIOSH do not constitute NIOSH endorsement of the sponsoring organizations or their programs or products. Furthermore, NIOSH is not responsible for the content of these Web sites. All web addresses referenced in this document were accessible as of the publication date.

The views expressed by non-NIOSH authors in these proceedings are not necessarily those of NIOSH.

Ordering Information

To receive documents or other information about occupational safety and health topics, contact NIOSH at

>Telephone: 1–800–CDC–INFO (1–800–232–4636)
>TTY: 1–888–232–6348
>e-mail: cdcinfo@cdc.gov

or visit the NIOSH Web site at www.cdc.gov/niosh.

For a monthly update on news at NIOSH, subscribe to NIOSH *eNews* by visiting **www.cdc.gov/niosh/eNews**.

DHHS (NIOSH) Publication No. 2009–145

June 2009

SAFER • HEALTHIER • PEOPLE™

Table of Contents

	Page
Foreword	iii
Acknowledgments	iv
Conference Organizing Committee	iv
Human Body Vibration Scientific Committee	iv
Introduction	1
Keynote Speakers	5

Podium Presentations

Session I: Epidemiology and Standards I	7
Session II: New Technologies in Human Vibration	14
Session III: Whole-Body Vibration I	25
Session IV: Human Body Modeling and Vibration I	36
Session V: Whole-Body Vibration II	47
Session VI: Human Body Modeling and Vibration II	56
Session VII: Hand-Arm Vibration	67
Session VIII: Health Effects	80
Session IX: Epidemiology and Standards II	91
Session X: Whole-Body Vibration III	110
Poster Session	121
Index of Authors	136

Foreword

We are proud to have cosponsored the Second American Conference on Human Vibration in Chicago, Illinois. Much hard work and diligent effort went into making this meeting a success and an avenue for future work in human vibration and health initiatives. The goal and the main thrust of the conference were to provide a forum for scientists, engineers, medical doctors, industrial hygienists, and educators to learn and advance research/education in the unique area of human body vibration. In promoting health and safety and in stimulating progress, leaders in the field were invited to share their insight and expertise in addition to the excellent and plausible papers on the presentation schedule.

We hope that the proceedings of the conference, published here, will serve as a means of continuing the dialogue.

This unique forum afforded participants opportunities to learn firsthand what their peers and colleagues are working on and to exchange information on a variety of relevant topics including human response, human modeling, experimental design, sensors, new technologies, and epidemiology studies in human responses to hand-transmitted and whole-body vibration. This research is essential for better understanding the risk factors for adverse effects related to vibration and for designing more effective interventions to prevent painful and potentially disabling work-related injuries.

This conference addressed contemporary issues regarding occupational health, prevention measures, and scientific data collection used to study the complex, dynamic human response to vibration. The agenda included a rich and diverse scientific program as researchers and medical professionals from around the world gathered to examine human responses to hand-transmitted vibration and whole-body vibration.

Special thanks are extended to the University of Illinois at Chicago for hosting the Second American Conference on Human Vibration as well as the scientific presenters from both the U.S. and abroad who shared their work and participated in advancing the science toward achieving safer and healthier workplaces.

Christine M. Branche, Ph.D.
Acting Director
National Institute for Occupational Safety and Health
Centers for Disease Control and Prevention

ACKNOWLEDGMENTS

The success of the Second American Conference on Human Vibration (2nd ACHV) could not be achieved except by the contributions of many. The conference Chair and Co-chair Farid Amirouche and Alan Mayton express their thanks and appreciation to the following individuals and organizations: the keynote speakers – Dr. Anthony Brammer, Mr. Thomas A. Broderick, Dr. Frank Buczek, Jr. (on behalf of Dr. John Howard), Dr. Mark Gonzalez, Dr. Rosemary Sokas; all of the reviewers who assisted in the technical review process including Kristine Krajnak, Ren Dong, John Wu, and Oliver Wirth at the Health Effects Laboratory Division (HELD) of the National Institute for Occupational Safety and Health (NIOSH), Morgantown, WV, and the Human Body Vibration Scientific Committee; Robert J. Tuchman, CDC Writer-Editor Services Branch, for his valuable insights and outstanding support in the editing, organization, and assembly of the final document; David Caruso (PRL) for designing the document cover, conference announcement, and call for papers; and Jill Raufus (CDC-MASO) for her support in organizing the review process and assembling this document.

2nd ACHV Organizing Committee

Farid Amirouche[1]
Alan Mayton[2]
Christopher Jobes[2]
Bertrand Valero[1]
Kimberly Balogh[1]
Urszula Sas[1]

[1]Department of Mechanical and Industrial Engineering, University of Illinois at Chicago
[2]Pittsburgh Research Laboratory, National Institute for Occupational Safety and Health

Human Body Vibration Scientific Committee

Farid Amirouche (Chairman)	University of Illinois at Chicago
Thomas Armstrong	University of Michigan
Paul-Émile Boileau	IRSST, Canada
Anthony Brammer	National Research Council, Canada
Martin Cherniack	University of Connecticut
Ren Dong	NIOSH – HELD
Thomas Jetzer	Occupational Medicine Consultants
Bernard Martin	University of Michigan
Robert Radwin	University of Wisconsin, Madison
Subhash Rakheja	Concordia University, Canada
Douglas Reynolds	University of Nevada, Las Vegas
Danny Riley	Medical College of Wisconsin
Suzanne Smith	U.S. Air Force Research Laboratory
Donald Wasserman	Human Vibration Consultant
Jack Wasserman	University of Tennessee, Nashville
David Wilder	University of Iowa

ACRONYMS AND ABBREVIATIONS

1-D	one-dimensional
AAALAC	Association for Assessment and Accreditation of Laboratory Animal Care
ACGIH	American Conference of Governmental Industrial Hygienists
ADAMS	Automatic Dynamic Analysis of Mechanical Systems
ANCOVA	analysis of covariance
ANOVA	analysis of variance
ANSI	American National Standards Institute
APMS	apparent mass
ASSE	American Society of Safety Engineers
BMD	bone mineral density
BMI	body mass index
BPF	blade passage frequency
BS	British Standard
CCM	compensatory control model
CDC	Centers for Disease Control and Prevention
CPT	current perception threshold
DPOAE	distortion product otoacoustic emission
EAV	exposure action value
ECM	extracellular matrix
ELV	exposure limit value
EMG	electromyography, electromyogram
EU	European Union
FAI	fast-adapting type I (receptor)
FBF	finger blood flow
FEA	finite-element analysis
fMRI	functional magnetic resonance imaging
FST	finger skin temperature
GPS	global positioning system
HAV	hand-arm vibration
HAVS	hand-arm vibration syndrome
HGCZ	health guidance caution zone
IFF	interstitial fluid flow
IRSST	Institut de recherche Robert-Sauvé en santé et en sécurité du travail (Occupational Health and Safety Research Institute Robert-Sauvé, Montreal, Quebec, Canada)
ISO	International Organization for Standardization (Geneva, Switzerland)
LHD	load-haul-dump
MSD	musculoskeletal disorder
MTVV	maximum transient vibration value
NAICS	North American Industry Classification System
NASA	National Aeronautics and Space Administration
NCV	nerve conduction velocity
NE	norepinephrine

NIHL	noise-induced hearing loss
NIOSH	National Institute for Occupational Safety and Health
NSERC	Natural Sciences and Engineering Research Council (Canada)
OEL	occupational exposure limit
Pabs	power absorbed
PDA	personal digital assistant
PEMF	pulsed electromagnetic field
PSD	power spectral density
PVC	polyvinyl chloride
RDS	response dose spectrum
rms	root-mean-square
ROS	reactive oxygen species
RS	response spectrum
rss	root-sum-of-squares
S&H	safety and health
SAI	slow-adapting type I (receptor)
SD	standard deviation
SDOF	single degree of freedom
SDOFS	single degree of freedom system
SEM	standard error of the mean
SMC	smooth muscle cell
SPL	sound pressure level
STHT	seat-to-head transmissibility
SUV	sport utility vehicle
TTS	temporary threshold shift
TWA	time-weighted average
UIC	University of Illinois at Chicago
VDV	vibration dose value
VPA	vibration power absorption
VPT	vibrotactile perception threshold
WBV	whole-body vibration
WBVT	whole-body vibration therapy
WFD	white finger disease

UNIT OF MEASURE ABBREVIATIONS

cm	centimeter
cm/ms	centimeters per millisecond
dB	decibel
dBA	decibel, A-weighted
dbNA	decibel nível de audição
df	degree of freedom
g	gram
g	acceleration due to gravity (1 g = 9.80665 m/s^2)
G	gauss
GPa	gigapascal
hr	hour
Hz	hertz
in	inch
kg	kilogram
kHz	kilohertz
km	kilometer
km/hr	kilometers per hour
kN	kilonewton
m	meter
m^3	cubic meter
m/s	meter per second
m/s^2, ms^{-2}, m/s/s	meter per second squared
mA	milliampere
mg/kg	milligram per kilogram
mg^2	square milligram
min	minute
mL	milliliter
mm	millimeter
mM	millimolar
mm Hg	millimeters of mercury
MPa	megapascal
mph	mile per hour
mt	metric ton
mV/cm	millivolt per centimeter
N	newton
nm	nanometer
Nm	net moment
N/m^2	newton per square meter
psi	pound-force per square inch
rad/s/s	radian per second squared
sec	second
μm	micrometer
μM	micromolar
μM^2	square micromolar
μT	microtesla
°C	degree Celsius

INTRODUCTION

There is a saying, "If it moves, it vibrates." This is especially true in our modern industrial environments, where workers use powered tools, machinery, vehicles, and heavy equipment. How workers are affected by these elements of the industrial working environment is a concern of health professionals, governments, and scientists around the world. How to adequately assess the effects of human vibration exposure is an equally important issue.

The human body comprises a dynamic structure that is living, intelligent, and complex. Consequently, it is not unreasonable to consider that exposing the body to an array of vibration environments would result in outcomes that are not necessarily simple or easily predictable. Sensations caused by vibration may include nausea, annoyance, discomfort or pain, and exhilaration or pleasure. Many factors contribute to the effects of vibration on humans, including the characteristics of the motion and the exposed individual, the activities of the exposed individual, and various elements of the environment. There have been attempts to summarize the effects of occupational vibration exposures by simply recommending the avoidance of certain frequencies of vibration or by providing a graphical method, i.e., a single curve displaying all human responses to all frequencies. However, such efforts have not withstood the scrutiny of scientific research and analysis. Thus, attendees and participants at this professional gathering now have the rare opportunity to benefit from the technological advances and new scientific research to enhance their understanding of human responses to vibration that suggest more than a simplistic approach.

In the tradition of the First American Conference on Human Vibration (1st ACHV) held in Morgantown, WV, June 2006, the 2nd ACHV provides a forum for scientists, engineers, medical doctors, industrial hygienists, and educators to learn and advance research and education in the specialized area of human body vibration. The conference was organized by the Mechanical and Industrial Engineering Department, University of Illinois at Chicago, with support from the National Institute for Occupational Safety and Health (NIOSH), Pittsburgh Research Laboratory. As with the first conference, the 2nd ACHV provides the unique opportunity to learn about the current activities of peers and colleagues and to exchange information on topics of interest such as human response, human modeling, experimental design, sensors, new technologies, and epidemiological studies of human responses to hand-transmitted and whole-body vibration. The conference organizers hope that the publication of these conference proceedings continues to foster new research and technological advances to reduce health hazards associated with occupational vibration exposures.

Farid Amirouche
Conference Chair

Alan G. Mayton
Conference Co-Chair

Outline of the Proceedings[1]

A conceptual model of the major factors influencing human vibration exposure and health effects, as well as their general relationships, is shown in Figure 1. Most of the studies presented at this conference involved the measurement, understanding, and/or assessment of these factors and their relationships. Other studies focused on the control or mitigation of vibration transmitted to the human body or the biological responses of tissues to vibration exposure. Therefore, the proceedings are organized in terms of the major factors and their relationships.

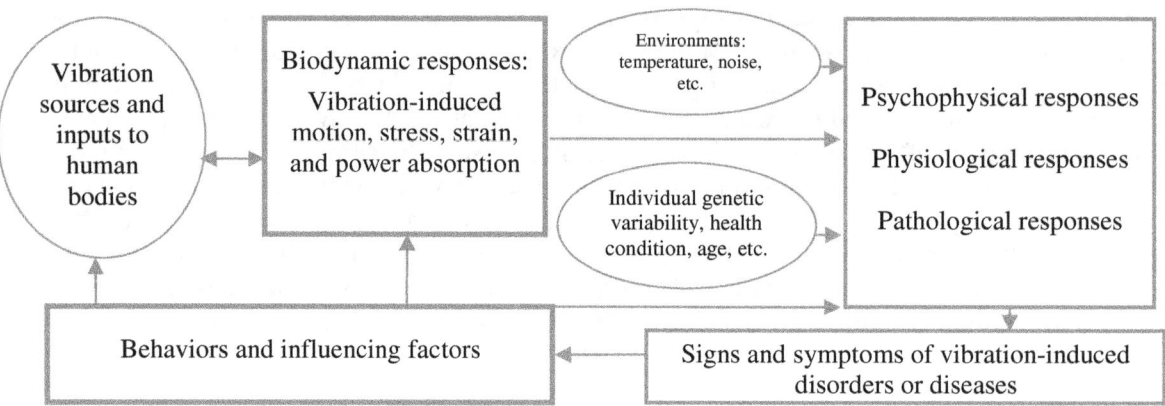

Figure 1.—Conceptual model of human vibration exposure and health effects.

Specific sessions focused on the measurement of whole-body (III, V, X) and hand-transmitted (VII) vibration in field and laboratory settings. Studies presented in other sessions (I, VIII, and IX) characterized the physiological responses of humans and animals to vibration and described how vibration may act on the cellular and molecular levels to induce these physiological changes. Studies describing the development and use of mechanical models for assessing the effects of vibration on the human body were presented in Session II. A number of presentations throughout this conference also focused on the assessment of current standards; presented data can be used to update and improve those standards. Together, the presentations in these 10 sessions covered state-of-the-art approaches for measuring, assessing, and mitigating the effects of human exposures to occupational vibration.

Assessment of the State of Human Body Vibration Research

To aptly assess the state of human body vibration research, we refer back to the insights of Michael J. Griffin, a keynote speaker at the 1st ACHV. Dr. Griffin sought to encapsulate this field of study in terms of "what we know that we know, what it is sometimes claimed that we know, and what we know that we do not know about the relation between exposures to vibration and our health." He speculated also on "what we do not know that we do not know" [Griffin 2006].

Griffin [2006] explains that "[t]here are many unknowns in the field of human responses to vibration. Not all would agree on what is known and what is unknown." With regard to hand-transmitted vibration, Griffin states: "We know that exposures to hand-transmitted vibration result in various disorders of the

---

[1] This section was compiled by Ren G. Dong, Kristine Krajnak, and Thomas W. McDowell of the NIOSH Health Effects Laboratory Division, Morgantown, WV.

hand, including abnormal vascular and neurological function. Not all frequencies, or magnitudes, or durations of hand-transmitted vibration cause the same effects."

We may claim to know that:

> To enable exposures to be reported and compared, they are "measured" and "evaluated" using defined (e.g., standardized) procedures. This involves identifying what is to be measured and specifying how it is expressed by one (or a few) numbers. Summarizing a vibration exposure in a single value involves assuming the relative importance of components within the vibration (e.g., different magnitudes, frequencies, directions, and durations), so standards define "weightings" for these variables. The importance of the weighted values may also be suggested, allowing "assessments" according to a criterion (e.g., the probability of a specific severity of a specific disease).

> Standards for the measurement and evaluation of hand-transmitted vibration define a frequency weighting and time dependencies that allow the severity of vibration exposures to be assessed and the probability of finger blanching to be predicted. [Griffin 2006]

What we do not know is that:

> [T]he frequency weighting in current standards reflects the relative importance of different frequencies and axes of vibration in producing any specific disorder. We do not know whether the energy-based daily time-dependency inherent in $A(8)$ reflects the relative importance of vibration magnitude and daily exposure duration. Consequently, the relation between $A(8)$ and the years of exposure to develop finger blanching, as in an appendix to ISO 5349-1 (2001), is not well-founded.

> We do not know, or at least there is no consensus on, the full extent of the disorders caused by hand-transmitted vibration (e.g., vascular, neurological, muscular, articular, central), or the pathogenesis of any specific disorder caused by hand-transmitted vibration, or the roles of other factors (e.g., ergonomic factors, environmental factors, or individual factors). We know that acute exposures to hand-transmitted vibration cause both vascular and neurological changes analogous to the changes seen in those occupationally exposed to hand-transmitted vibration, but we do not yet know how the acute changes relate to the chronic disorders. [Griffin 2006]

Regarding whole-body vibration, Griffin [2006] reports: "We know that many persons experience back pain and that some of these are exposed to whole-body vibration. We know that in the population at large, occupational exposures to whole-body vibration are not the main cause of back problems and that ergonomic factors (e.g., lifting and twisting) and personal factors are often involved. We know vibration and shock can impose stresses that could supplement other stresses." We may claim to know that "[m]easurement methods and evaluation methods have been defined in which the frequencies, directions, and durations are weighted so as to predict the relative severity of different vibrations and indicate the magnitudes that might be hazardous." What we do not know is that "[w]e are not able to predict the probability of any disorder from the severity of an exposure to whole-body vibration. We do not know whether there is any disorder specific to whole-body vibration or what disorders are aggravated by exposure to whole-body vibration. We do not know the relative importance of vibration and other risk factors in the development of back disorders."

Griffin [2006] concludes that:

> Providing guidance to others involves compromises—a perceived need, or other argument, that may outweigh the cautious interpretation of scientific evidence. Standards for measuring and evaluating human exposures to vibration use uncertain frequency weightings and time dependencies but allow legislation for the protection of those exposed. The standards may appear useful, but it is prudent to

distinguish between standards and knowledge—between what is accepted to reach a consensus and what can be accepted as proven. Standards may guide actions but not understanding.

Where reducing risk solely involves reducing vibration magnitude or exposure duration, ill-founded evaluation methods will not increase risk. Where prevention involves a redistribution of vibration over frequencies or directions, or balancing a change in magnitude with a change in duration, an inappropriate evaluation method can increase risk. For example, the hand-transmitted vibration frequency weighting, which may be far from optimum, implies that gloves give little beneficial attenuation, whereas a different weighting might indicate that gloves can be a useful means of protection.

What do we not know that we need to know? Not all appreciate the benefits of placing more reliance on traceable data than on consensus. Traceability is fundamental to quality systems but deficient in current standardization. Standards can comfort their users—justifying actions without resort to understanding—while concealing assumptions that may prevent the minimization of the risks of injury from exposures to vibration.

Reference

Griffin MJ [2006]. Health effects of vibration: the known and the unknown. In: Proceedings of the First American Conference on Human Vibration. Morgantown, WV: U.S. Department of Health and Human Services, Centers for Disease Control and Prevention, National Institute for Occupational Safety and Health, DHHS (NIOSH) Publication No. 2006–140, pp. 3–4.

KEYNOTE SPEAKERS

Frank Buczek, Jr. (on behalf of Dr. John Howard, Director, National Institute for Occupational Safety and Health (NIOSH), Washington, DC)

Frank L. Buczek, Jr., Ph.D., is Chief of the Engineering & Control Technology Branch, in the Health Effects Laboratory Division, of the National Institute for Occupational Safety & Health in Morgantown, WV. He also serves as Coordinator for the Musculoskeletal Disorders (MSD) Cross Sector Program at NIOSH. He holds adjunct appointments at several universities, is a Past President of the Gait and Clinical Movement Analysis Society, and serves on MSD-related study sections for the National Institutes of Health (NIH). Building upon an undergraduate foundation in Mechanical Engineering, Dr. Buczek earned his doctorate in Human Movement Biomechanics from the Pennsylvania State University in 1990, where early research efforts challenged slip resistance levels considered adequate for flooring at that time. During postdoctoral work at NIH, he specialized in advanced kinematic and kinetic modeling of anatomical joints. Since 1995, he has applied this expertise to pediatric gait analysis, leading most recently to collaboration on muscle actuated, full-body simulations of normal and pathological gait.

Mark Gonzalez

Mark Gonzalez, M.D., is a Professor at the University of Illinois at Chicago (UIC), Department of Orthopedic Surgery. Mark has been a Fellow of the American Academy of Orthopedic Surgeons since 1993 and is Chairman of Orthopedic Surgery, Cook County Hospital, Chicago, IL. Mark received his B.S. in biochemistry from the University of Illinois-Champaign in 1976 and his M.D. from UIC in 1980. He completed his orthopedic residency at UIC during 1981–1985 and fellowships at Ohio State University and the University of Louisville. He is currently the Chairman of the Department of Orthopedic Surgery and is completing his Ph.D. in mechanical engineering at UIC. He is the author of more than 100 technical papers and has been a Fellow of the American Academy of Orthopedic Surgeons since 1992. His current research includes the areas of nerve regeneration, orthopedic biomechanics, and micro-electro-mechanical system (MEMS) sensor devices.

Anthony Brammer

Anthony (Tony) Brammer, Ph.D., is a Professor of Medicine at the University of Connecticut Health Center, and works in the Ergonomic Technology Center of Connecticut. He is also a Visiting Scientist at the National Research Council of Canada, where he has conducted research in acoustics for most of his career, and an Associate of Envir-O-Health Solutions, which is an Ottawa company specializing in solutions to environmental and occupational health problems. Dr. Brammer received a National Research Council Research Fellowship in 1968, the National Research Council's Achievement Award in 1994, and the National Research Council's Institute for Microstructural Sciences' Award in 2001, and the International Commission on Occupational Health's Service Award in 2006. He is a Fellow of the Royal Society of Medicine and of the Acoustical Society of America, and is past Chairman of the Scientific Committee on Vibration and Noise of the International Commission on Occupational Health. He is also Convener (Chairman) of Working Groups of the International Organization for Standardization (ISO) on vibrotactile perception and biodynamic modeling.

Thomas A. Broderick

Thomas A. Broderick, M.S., is Executive Director, Construction Safety Council, Chicago, IL. Tom has been involved in the construction field since the early 1970s. He has been a construction worker, co-owner of a small construction company, and a construction safety professional. During his safety career, he has managed safety programs at projects including nuclear and fossil fuel power plants, paper mills, and other large jobs for contractors such as Blount Construction Co., Stone & Webster Engineering Corp., and the Rust Engineering Co. Tom holds a B.S. degree in speech communications and a master's degree in safety management. He is currently the executive director of the not-for-profit Chicagoland Construction Safety Council and the Construction Safety Council. In that capacity he is the director of the annual Construction Safety Conference in Chicago, the largest educational gathering in the United States focused solely on construction environmental health and safety (EHS) issues. Tom was nominated in 2001 by Secretary of Labor Elaine Chao to serve on the congressionally mandated Advisory Committee for Construction Safety and Health, a committee on which he still serves. He is also past president of Veterans of Safety International and a commissioner on the National Commission for the Certification of Crane Operators. He has authored numerous articles on EHS and has contributed to several texts in the field.

Rosemary Sokas

Rosemary Sokas, M.D., M.S., is a Professor of Environmental and Occupational Health Sciences (EOHS), University of Illinois at Chicago. Her academic background is: M.D., Boston University, 1974; M.S., Physiology, Harvard University, 1981; MSPH, Harvard University, 1980; and BSMLS, Medical Science, Boston University, 1974. Her research interests include applied, translational occupational safety and health organizational work targeting small businesses and vulnerable populations. Her teaching interests include problem-based occupational and environmental health/integration of EOHS into public health and primary care.

Session I: Epidemiology and Standards I

Chair: Kristine Krajnak
Co-Chair: Thomas McDowell

Presenter	Title	Page
R. House University of Toronto	Current Perception Threshold, Nerve Conduction Studies, and the Stockholm Sensorineural Scale in Workers With Hand-Arm Vibration Syndrome	8
S. Govindaraju Medical College of Wisconsin	Mechanism of Vibration-Induced Smooth Muscle Cell Injury in the Rat Tail Artery	10
K. Krajnak National Institute for Occupational Safety and Health (NIOSH)	Sensory Nerve Responses to Acute Vibration Are Frequency-Dependent in a Rat Tail Model of Hand-Arm Vibration Syndrome	12

CURRENT PERCEPTION THRESHOLD, NERVE CONDUCTION STUDIES, AND THE STOCKHOLM SENSORINEURAL SCALE IN WORKERS WITH HAND-ARM VIBRATION SYNDROME

Ron House,[1] Kristine Krajnak,[2] and Michael Manno[1]

[1]University of Toronto and St. Michael's Hospital, Toronto, Ontario, Canada
[2]Health Effects Laboratory Division, National Institute for Occupational Safety and Health, Morgantown, WV

Introduction

Neurological abnormalities are frequently reported in workers using vibrating tools. The Stockholm sensorineural scale attempts to classify the neurological component of hand-arm vibration syndrome. However, it is unclear what the underlying neurological damage consists of, how it should be measured, and how it relates to the Stockholm scale.

This study was carried out to examine the relation between the Stockholm sensorineural scale and tests of neurological abnormalities, including measurement of current perception threshold (CPT) and nerve conduction in workers using vibrating tools.

Methods

All of the study participants had been exposed to hand-arm vibration at work and were assessed at St. Michael's Hospital, Toronto, Ontario, Canada, over a 4-year period ending in 2003. The assessment consisted of an occupational and medical history, physical examination with emphasis on the neurological examination of the upper extremities, and neurological tests including measurement of CPT and nerve conduction. The results of the medical history and physical examination were used to determine the Stockholm sensorineural scale. The CPT measurements, nerve conduction measurements, and determination of the Stockholm scale were all carried out in a fashion blinded to each other.

Polychotomous logistic regression was used to examine the relation between the Stockholm scale dependent variable and various independent variables including (1) the CPT results for the median and ulnar nerve at 5250 and 2000 Hz and (2) the presence of neuropathy on the nerve conduction study (median, ulnar, digital).

Results

There were 155 study participants who worked in a variety of industries, with mining being the most prevalent. The workers were all male and had an average age of 45.9 ± 11.32 years and an average duration of exposure to vibrating tools of 21.8 ± 12.03 years. The distribution of subjects in the Stockholm scale categories 0, 1, ≥ 2 was as follows: right hand – 51, 81, 23; left hand – 52, 83, 20. The nerve conduction studies indicated that 40.6% of subjects had peripheral neuropathy in the right hand (34.8% median; 6.5% ulnar, 1.3% digital) and 30.3% had peripheral neuropathy in the left hand (23.9% median; 6.5% ulnar, 1.3% digital). The sum of the percentages of the specific neuropathies exceeded the overall total percentage of subjects with peripheral neuropathy in each hand because some subjects had more than one neuropathy. None of the CPT measurements at any frequency was found to be associated with any type of neuropathy. The polychotomous logistic regression results indicated that the CPT measurements at 2000 Hz were the main predictor variables for the Stockholm scale. In the right hand, the only statistically significant independent variable was the CPT median nerve value at 2000 Hz ($p < 0.0001$). In the left hand, the main statistically significant independent variables were the CPT median nerve value at 2000 Hz ($p = 0.001$) and the CPT ulnar nerve value at 2000 Hz ($p = 0.0007$). None of the nerve conduction abnormalities was significantly associated with the Stockholm scale in the regression results in either hand.

Discussion

The results indicated that the main neurological test predictive of the Stockholm scale was the CPT measurement at 2000 Hz. This independent variable measures damage to the large myelinated fibers of the fingers.

A previous study by Lander et al. [2007] in the same setting but using different data suggested an association between overall CPT results for the ulnar nerve and the Stockholm scale. The results presented here provide more detailed information about the relation between frequency-specific CPT measurements and the Stockholm scale. Our findings are similar to those described by Kurozawa and Nasu [2001].

There is also evidence of similar effects on CPT in rats exposed to vibration [Krajnak et al. 2007], as well as histological evidence of damage to large myelinated fibers due to vibration exposure in animals [Chang et al. 1994]. These findings provide biological plausibility for damage to the large myelinated fibers in workers exposed to vibration and suggest overall coherence of evidence at the histological, animal, and human level.

Workers exposed to vibration also have a high prevalence of neuropathy proximal to the hand. However, our results indicated that these proximal neuropathies were not associated with the Stockholm scale. Also, the proximal lesions did not explain the CPT findings in our study. Therefore, the CPT abnormalities, which were clearly associated with the Stockholm scale, seem to indicate a second type of distal neurological lesion in the fingers due to vibration exposure.

Further research should focus on the best methods of measurement and the functional significance of these distal neurological lesions. In addition, the role of vibration and ergonomic factors in the etiology of the proximal neuropathies needs to be understood in more detail. It is also evident that improvement is needed in the classification system used for the neurological damage associated with vibration. Improved understanding in these areas should lead to enhanced prevention efforts.

References

Chang KY, Ho ST, Yu HS [1994]. Vibration induced neurophysiological and electron microscopy changes in rat peripheral nerves. Occup Environ Med *51*(2):130–135.

Krajnak K, Waugh S, Wirth O, Kashon ML [2007]. Acute vibration reduces Abeta nerve fiber sensitivity and alters gene expression in the ventral tail nerves of rats. Muscle Nerve *36*(2): 197–205.

Kurozawa Y, Nasu Y [2001]. Current perception thresholds in vibration-induced neuropathy. Arch Environ Health *56*(3):254–256.

Lander L, Lou W, House R [2007]. Nerve conduction studies and current perception thresholds in workers assessed for hand-arm vibration syndrome. Occup Med (Lond) *57*(4):284–289.

MECHANISM OF VIBRATION-INDUCED SMOOTH MUSCLE CELL INJURY IN THE RAT TAIL ARTERY

Sandya Govindaraju, James Bain, and Danny Riley
Department of Cell Biology, Neurobiology, and Anatomy, Medical College of Wisconsin, Milwaukee, WI

Introduction

Vasospastic blanching of the fingers is a major complication of hand-arm vibration syndrome, an occupational disorder in workers using handheld power tools. Our rat tail vibration model, which simulates hand-transmitted vibration, was developed to investigate the cellular mechanism of vibration injury [Curry et al. 2002]. Vibration causes vasoconstriction and smooth muscle cell (SMC) vacuoles in the rat tail artery [Curry et al. 2005]. Vibration injury of SMC in arteries is hypothesized as a two-step process (Figure 1). The first step is SMC contraction and vacuole formation, stimulated by vibration via a centrally mediated, somatosympathetic neural response. The second step occurs when the protruding vacuoles are detached from the cell by the mechanical force of vibration. Detachment could generate persistent vacuoles on the one hand and increased fragmentation on the other. Other vacuoles may be resorbed by the parent cell during SMC relaxation. The number and size of vacuoles formed and lost will determine the severity of vascular damage induced by vibration. The present study addressed two questions: (1) Do vibration-induced vasoconstriction and vacuole formation occur in SMC of rats anesthetized to blunt central nervous system vasomotor activity? (2) Does vibration of anesthetized rats disrupt vacuoles generated by norepinephrine (NE)-induced vasoconstriction pretreatment?

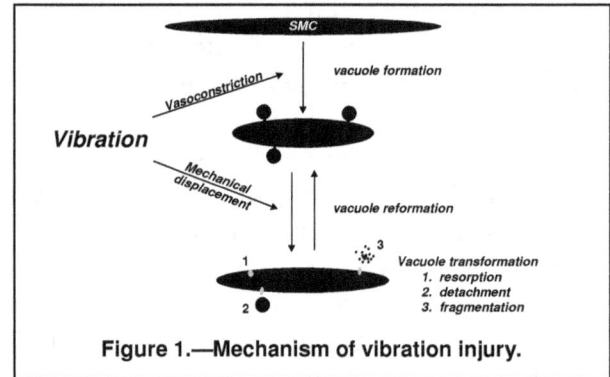

Figure 1.—Mechanism of vibration injury.

Methods

Awake rats were restrained on a nonvibrating platform with their tails taped to a vibrating stage. Vibration consisted of linear vertical oscillations with a frequency of 60 Hz, peak-to-peak amplitude of 0.98 mm, and an acceleration of 49 m/s^2 (rms) for 4 hr. Sham vibration rats were treated similarly but not vibrated. NE (1 mM) in Hanks Balanced Salt Solution bathed for 15 min the caudal artery exposed by a ~5-cm ventral skin incision. Following vibration, the tail segment C7 was removed and immersion-fixed in 4% glutaraldehyde, 2% paraformaldehyde in cacodylate buffer (pH 7.4). Arteries were postfixed in 1.3% osmium tetraoxide and embedded in epoxy resin for semithin (0.5 μm) and ultrathin (~70 nm) sectioning. Version 1.28v ImageJ software was used to count vacuoles, measure lumen circumference, and determine internal elastic membrane length in toluidine blue-stained semithin sections. Vacuoles counted were round to oval-shaped intracellular inclusions (2–12 μm), clear or homogeneously dense material.

Results

Vasoconstriction and vacuole formation occurred following awake-vibration and NE treatment. Vacuole number remained high 4 hr after exposure to NE (Table 1).

Compared to the awake-sham group, the NE + anesthesia-vibration group had larger areas of dense filamentous matrix (Figure 2). Compared to awake-sham, the awake-vibration and NE-anesthesia-vibration groups had a higher content of cellular membranous and particulate debris. The cellular debris in the NE + anesthesia-vibration group was larger than that of the awake-vibration group.

Table 1.—Lumen size[1] and vacuole number

Treatment	No. of rats per group	Lumen	SMC vacuoles[2]
Awake-sham	7	52.3 ± 3.8	3.6 ± 2.3
Awake-vibration	7	[3]42.1 ± 3.6	[4]44.8 ± 6.1
Anesthesia-sham	6	54.5 ± 3.3	2.0 ± 1.0
Anesthesia-vibration	6	53.4 ± 3.1	4.0 ± 1.0
Vehicle application	5	59.8 ± 3.2	0.0 ± 0.0
NE	7	[3]29.4 ± 2.0	[3]106.6 ± 16.3
NE + anesthesia-sham	6	61.0 ± 2.7	[4]44.5 ± 8.3
NE + anesthesia-vibration	8	52.0 ± 4.2	0.4 ± 0.2

[1]Percent ratio of the lumen circumference to the internal elastic membrane length.
[2]Total numbers per section.
[3]Significantly different from all other groups.
[4]Significantly different ($p<0.05$) from all other groups, except each other, when tested using the Student-Newman-Keuls pairwise comparison test.

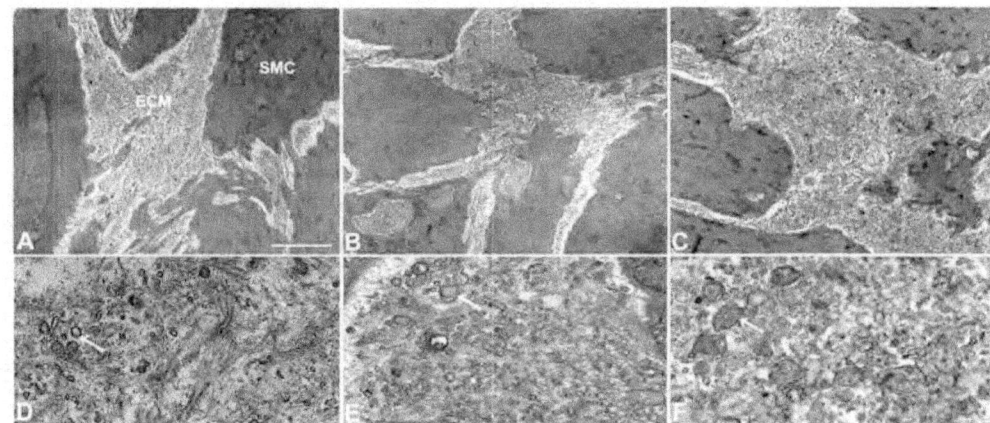

Figure 2.—Electron micrographs of extracellular matrix (ECM) of the awake-sham (A and D), awake-vibration (B and E), and NE + anesthesia-vibration (C and F) groups. Bar in A equals 3.6 μm for A through C and 440 nm for D through F.

Summary and Conclusions

- Anesthesia prevents vibration-induced vasoconstriction and vacuole formation.
- Vacuoles preformed by NE treatment are disrupted by vibration, as indicated by increased cellular debris in the ECM.
- Vibration causes cell injury in a two-step process: inducing vasoconstriction to generate vacuoles and disrupting vacuoles to cause loss of portions of SMC.
- Repeated cell injury over time may compromise vascular function and trigger adaptive cellular mechanisms to combat vibration injury.

References

Curry BD, Bain JLW, Yan J-G, Zhang LL, Yamaguchi M, Matloub HS, Riley DA [2002]. Vibration injury damages arterial endothelial cells. Muscle Nerve 25(4):527–534.

Curry BD, Govindaraju SR, Bain JL, Zhang LL, Yan JG, Matloub HS, Riley DA [2005]. Evidence for frequency-dependent arterial damage in vibrated rat tails. Anat Rec A Discov Mol Cell Evol Biol 284(2):511–521.

SENSORY NERVE RESPONSES TO ACUTE VIBRATION ARE FREQUENCY-DEPENDENT IN A RAT TAIL MODEL OF HAND-ARM VIBRATION SYNDROME

Kristine Krajnak,[1] Stacey Waugh,[1] G. Roger Miller,[1] and Michael L. Kashon[2]

[1]Engineering and Control Technology Branch, Health Effects Laboratory Division
[2]Biostatistics and Epidemiology Branch, Health Effects Laboratory Division

National Institute for Occupational Safety and Health (NIOSH), Morgantown, WV

Introduction

Occupational exposure to hand-arm vibration through the use of powered handtools can result in reductions in tactile sensitivity. Acute exposures to vibration cause shifts in vibrotactile thresholds in exposed fingers that are affected by the frequency of the exposure, with exposure to midrange vibration frequencies (i.e., around 125 Hz) producing the greatest shifts [Maeda and Griffin 1994]. We have demonstrated that acute vibration exposure at 125 Hz results in a transient increase in sensory nerve thresholds to a 2,000-Hz transcutaneous stimulus in a rat tail model of hand-arm vibration syndrome [Krajnak et al. 2007]. The goal of this study was to determine how repeated exposures to vibration at frequencies that produce different biodynamic responses affect sensory nerve function in our rat tail model.

Methods

Animals. Male Sprague Dawley rats (8 weeks of age, n = 5 per group) were housed in AAALAC-accredited facilities. All procedures were approved by the NIOSH Animal Care and Use Committee and were in compliance with CDC guidelines for the care and use of laboratory animals. Vibration exposures were performed by restraining rats in a Broome-style restrainer and securing their tails to a vibration platform using 6-mm-wide straps that were placed over the tail every 3 cm. Restraint control animals were treated in an identical manner except that the tail platform was set on isolation blocks instead of a shaker. Rats were exposed to 4-hr bouts of vibration or restraint for 10 consecutive days at 0, 62.5, 125, or 250 Hz with a constant acceleration of 49 m/s^2 root-mean-square. Previous work in our laboratory has shown that vibration exposure at these frequencies and acceleration produce different magnitudes of transmissibility to the tail [Welcome et al. 2006]. An additional group of animals served as cage control rats.

Current perception thresholds (CPTs) and mechanoreceptor sensitivity. Sensory nerve function was assessed by measuring CPTs with a Neurometer (Neurotron, Inc., Baltimore, MD). Transcutaneous nerve stimulation at 2,000 Hz was applied to the C10 region of the tail to determine the effects of vibration on Aβ nerve fiber sensitivity. The intensity of the stimulus was automatically increased in small increments until the rat flicked its tail. Aβ-nerve fibers are large-diameter, myelinated nerve fibers that carry information from mechanoreceptors to the central nervous system. Tests were repeated until the animals displayed two responses that were within 2 CPT (or 0.02 mA) of each other (two to three tests per animal). To determine if changes in sensory nerve responsiveness were accompanied by changes in mechanoreceptor sensitivity, the response to pressure induced by the application of 1- and 10-g von Frey filament was assessed. If animals flicked their tail, they were responsive; if not, they were nonresponsive. CPT and mechanoreceptor sensitivity tests were performed before (pretest) and immediately after (posttest) the vibration exposure on days 1 and 9 of the exposure.

Data analyses. CPT data were analyzed using a mixed model three-way (condition × days of exposure × pre/post exposure) ANOVA where animal served as a random variable. The number of animals responding to stimulation with the von Frey filaments was analyzed using contingency table chi-square tests. Differences with $p < 0.05$ were considered significant.

Results

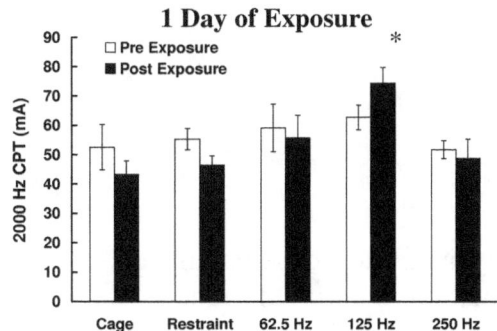
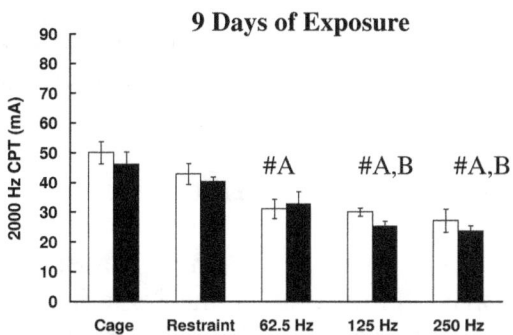

Figure 1.—CPTs after vibration exposure. After 1 day of exposure, vibration at 125 Hz resulted in an increase in the 2,000-Hz CPT threshold (*greater than preexposure, $p < 0.05$). After 9 days of exposure, CPTs were reduced in all rats exposed to vibration (# less than day 1, $p < 0.05$). In addition, vibrated animals displayed lower thresholds than control animals after 9 days of exposure (A: different from cage controls; B: different from restraint controls, $p < 0.05$).

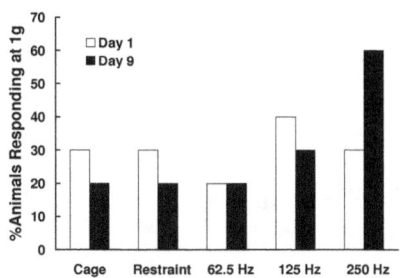

Figure 2.—Percentage of animals responding to mechanoreceptor stimulation with a 1-g von Frey filament after 1 or 9 days of vibration. The percentage of animals responding to stimulation with the filament was higher after 9 days of vibration exposure at 250 Hz than after 1 day (χ^2, $p < 0.04$). No changes were seen with 10-g stimulation.

Discussion

- As previously shown [Krajnak et al. 2007], a single exposure to vibration at 125 Hz results in an increase in the 2,000-Hz CPT. Similar responses to acute vibration are also seen in humans [Maeda and Griffin 1994].
- After 9 days of vibration exposure, 2,000-Hz thresholds were lower in vibrated than in control or restrained rats. In contrast, changes in mechanoreceptor sensitivity were seen only in rats exposed to 250 Hz, which is the frequency that generates the greatest biodynamic response in the tail [Welcome et al. 2006].
- Increased sensitivity to mechanical stimuli may serve as an early indicator of vibration-induced nerve damage.

References

Krajnak K, Waugh S, Wirth O, Kashon ML [2007]. Acute vibration reduces Abeta nerve fiber sensitivity and alters gene expression in the ventral tail nerves of rats. Muscle Nerve *36*(2): 197–205.

Maeda S, Griffin MJ [1994]. Temporary threshold shifts in fingertip vibratory sensation from hand-transmitted vibration and repetitive shock. Nagoya J Med Sci *57*(Suppl):185–198.

Welcome D, Dong RG, Krajnak KM [2006]. A pilot study of the transmissibility of the rat tail compared to that of the human finger. In: Proceedings of the First American Conference on Human Vibration (June 5–7, 2006). Morgantown, WV: U.S. Department of Health and Human Services, Public Health Service, Centers for Disease Control and Prevention, National Institute for Occupational Safety and Health, DHHS (NIOSH) Publication No. 2006–140, pp. 101–102.

Session II: New Technologies in Human Vibration

Chair: Michael Griffin
Co-Chair: Tom Jetzer *

Presenter	Title	Page
N. Shibata Japan National Institute of Occupational Safety and Health	Comparison Between Biodynamic Response Parameters of the Same Subject Obtained From Two Different Vibration Systems	15
R. Stayner RMS Vibration Test Laboratory	Suspended Seats for Reducing Exposure to Whole-Body Vibration: Improving Standard Tests to Reduce Interlaboratory Variations	17
S. D. Smith Wright-Patterson Air Force Base	Multiaxis Seat Cushion Transmissibility Characteristics	19
F. Amirouche University of Illinois at Chicago	Comparison of the Performance of a Pneumatic Active Seat Suspension Using a Discrete and Fuzzy Logic-Based Control Law	21
P. Marcotte Institut de recherche Robert-Sauvé en santé et en sécurité du travail (IRSST)	Characterization of the Noise and Vibration Produced by Portable Power Tools Used in the Automotive Repair Industry	23

* Author unable to attend session.

COMPARISON BETWEEN BIODYNAMIC RESPONSE PARAMETERS OF THE SAME SUBJECT OBTAINED FROM TWO DIFFERENT VIBRATION SYSTEMS

N. Shibata and S. Maeda
Japan National Institute of Occupational Safety and Health, Kawasaki, Japan

Introduction

Biodynamic response data of the hand-arm system give us much information not only for understanding hand-transmitted vibration and its health effects, but also for designing low-vibration exposure handheld power tools. Although ISO 10068 (1998) specifies the free, mechanical impedance of the human hand-arm system at the driving point, this standard contains some recommendations partially based on unreliable data sets, which differ considerably from each other even if measured under similar experimental conditions. These differences are partly due to the difference in dynamic characteristics of the experimental systems used, which of course include handle geometry and its setup. In this study, data compatibility was examined for two hand-arm vibration systems—one installed at the Japan National Institute of Occupational Safety and Health, the other at the U.S. National Institute for Occupational Safety and Health (NIOSH)—by measuring biodynamic response parameters of the hand-arm system with the same subject.

Methods

The experiments were conducted in two laboratories: one at the U.S. NIOSH, the other at the Japan NIOSH. The configurations of the two hand-arm vibration systems were basically the same. These vibration exposure systems consisted of an electromechanical shaker, horizontally mounted on a solid base, which makes vibration in the Z_h direction. The same force plates (Kistler 9286AA) were used to measure the feed force applied through the subject's hand to instrumented handles connected to the shaft of the shaker. The handles, of the same design [Dong et al. 2004], were composed of the base and measuring cap between which two force sensors (Kistler 9212) were sandwiched along the handle centerline. The grip force was measured with these force sensors. Also, an accelerometer (PCB 339B24) was secured at the center point of the measuring cap to measure and control the vibration acceleration of the handles.

The experiments were performed with one healthy male subject (aged 40 years). The subject was a nonsmoker and had never been exposed to high levels or long periods of hand-arm vibration occupationally or in his leisure time activities. The experiments were approved by the Research Ethics Committee of the Japan NIOSH.

The vibration signal used in this study was a pseudorandom signal with a broadband spectrum ranging from 10 to 1250 Hz of a constant power spectrum density of 1.0 $(m/s^2)^2/Hz$. We prepared two force coupling conditions: grip forces of 30 N and 50 N with a push force of 0 N. For each force coupling condition, five trials were conducted.

The subject posture used in this study was based on ISO 10819 requirements. The subject controlled the grip and feed forces under a certain force coupling condition by watching monitors during the measurement.

Results and Discussion

Figure 1 shows biodynamic response parameters measured with the same subject under the two different vibration systems. For each force coupling condition, the apparent mass and mechanical impedance obtained from biodynamic response measurements for the two vibration systems showed good agreement with each other. Differences of the apparent mass magnitudes observed in relatively low frequency ranges (<40 Hz) at the palm were statistically tolerable. Our results suggest that biodynamic response data obtained from the hand-arm vibration system at the Japan NIOSH can be comparable with data obtained from the system at the U.S. NIOSH.

Acknowledgment: The authors are grateful to Ren Dong of NIOSH, Morgantown, WV, for his support to perform our experiments.

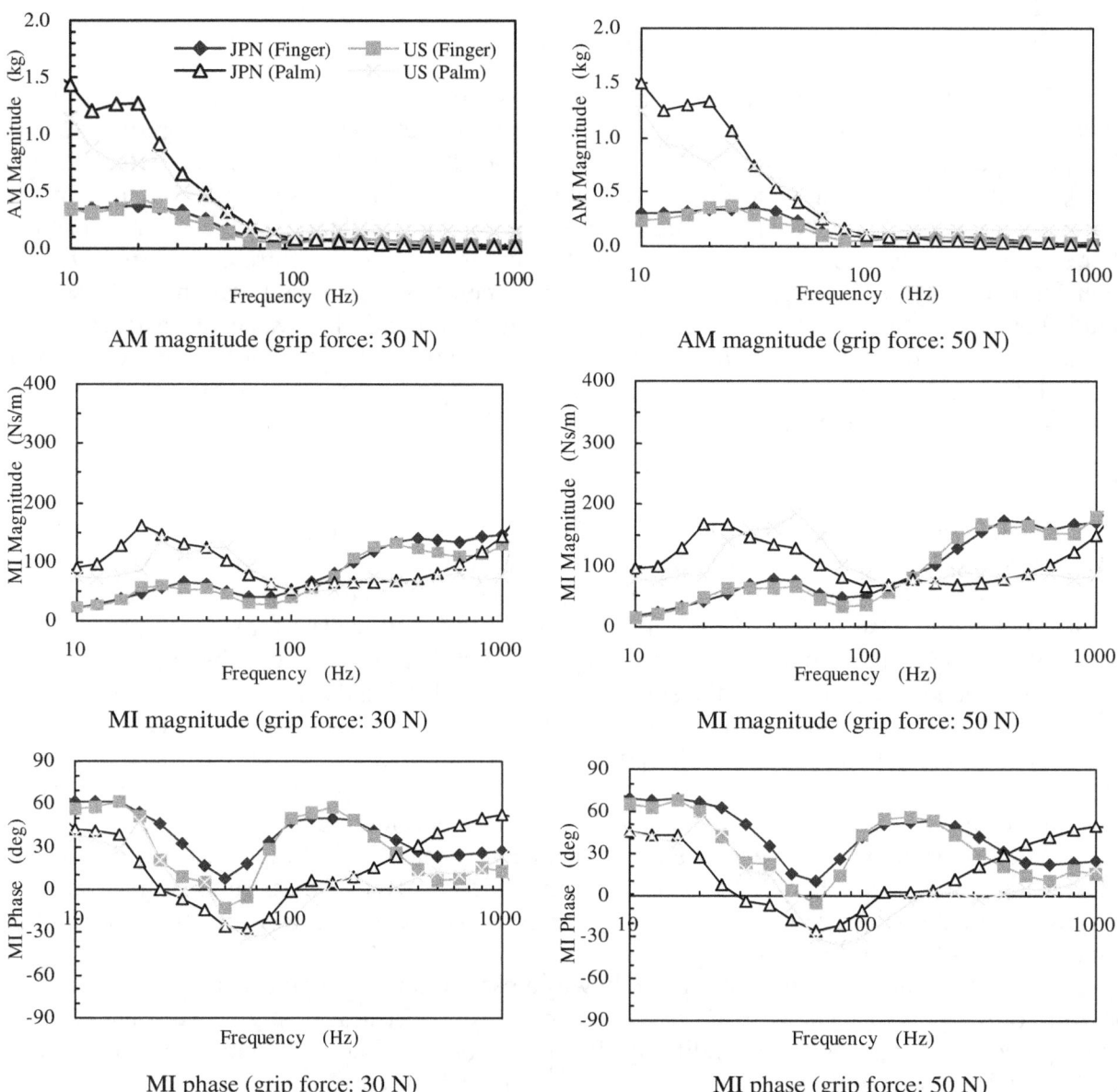

Figure 1.—Biodynamic response parameters given as functions of vibration frequency. (AM = apparent mass; MI = mechanical impedance).

References

Dong RG, Welcome DE, McDowell TW, Wu JZ [2004]. Biodynamic response of human fingers in a power grip subjected to a random vibration. J Biomech Eng *126*(4):447–457.

ISO [1998]. Mechanical vibration and shock – Free, mechanical impedance of the human hand-arm system at the driving point. Geneva, Switzerland: International Organization for Standardization. ISO 10068:1998.

SUSPENDED SEATS FOR REDUCING EXPOSURE TO WHOLE-BODY VIBRATION: IMPROVING STANDARD TESTS TO REDUCE INTERLABORATORY VARIATIONS

Richard Stayner
RMS Vibration Test Laboratory, Ludlow, U.K.

Introduction

Suspended seats are intended to reduce whole-body vibration transmitted from moving machines into the bodies of seated operators. The effectiveness of any particular seat suspension varies according to the application and operating conditions and depends on the frequency content and magnitude of the input vibration. At its simplest, the system operator-suspension acts like a mass on a spring, so that it can attenuate input vibrations at frequencies higher than $\sqrt{2}$ times the system natural frequency. In reality, damper characteristics and friction effects make the system nonlinear so that performance changes with the magnitude of the input, and large excursions bring into play the suspension end-stops. This complicates the selection of a suspension for a specific type of machine. Standard tests have been developed to overcome some of these difficulties. A critical part of these tests is to measure the vibration transmission of the seat for input vibration that is representative of the relevant machine type. The test codes specify the input vibrations and the weights of the operators (test persons). However, at least two attempts to conduct "round robin" tests among groups of European laboratories have shown differences of more than 25% in the measured performance of the same seats. CEN/TC231/WG9 (Seating), which is draft guidance on writing the test codes (EN 30326-1), is presently addressing ways to improve reproducibility, some of which are discussed below.

Setting Up the Seat: Ride Height and Subject Posture

With the help of the BGIA laboratory in Germany, WG9 ran a series of tests in which experienced test engineers from six laboratories measured the performance of two seats, each with an appropriate input vibration. For these tests, the engineers all used the same input vibrations and the same test persons. The "experimental variables" was the method used by each test engineer to determine the suspension height settings for the tests and the postures of the test persons. Surprisingly, in the worst case, the differences among test engineers exceeded 20%. From discussions with the engineers, a number of improvements to the guidance given in the test codes were suggested:

- Specify a common way to measure the suspension travel and determine the midride position to be used for vibration measurement.
- Display the suspension position continuously, first as an aid to correct setting, and then to allow corrections such as might be needed as the damper heats up.
- Specify posture better, in particular, control the pressure under the thighs of the test person to limit vibration transmitted to the seat cushion through the legs. Specific instructions may also be needed to control muscle tone.
- Specify test persons by sitting weight as opposed to standing weight.

Controlling the Input Vibration

In a second interlaboratory comparison, seven laboratories sent recordings of their input vibrations measured on the platform to an independent laboratory (IMMM, Slovak Republic) for comparison with the specifications, including tolerances on frequency spectra, and amplitude distributions. Again, there were some surprises. Some aspects of the specifications were clearly impractical and therefore were being ignored. In addition to correcting these points, the main advice was that the method of frequency analysis should be defined, i.e., common parameters should be used for fast Fourier transforms, such as sample rate (Δt), block length (Δf), time domain window, and amount of overlap. It should be noted that the differences found in both frequency and magnitude were small and were not experimentally correlated with any differences in measured vibration reduction.

Replacing the Human Subject

One obvious potential source of variation among laboratories is the difference in mechanical impedance (or apparent mass) of test persons, even when they have the same overall body mass. This, together with difficulties in finding very specific test persons, and aspects of subject safety have for years been behind attempts to replace live subjects with dummies. The frequency range of interest for suspended seats (vertical or z-axis suspension) extends from <1 Hz up to 20 Hz. The human body does not respond as a pure mass throughout all of that range; thus, many of the studies have been with dynamic dummies, starting with Suggs et al. [1969] and continuing up to the present with active models, e.g., Mozaffarin et al. [2007]. Nélisse et al. [2006] suggest that for applications with only very low-frequency components, a simple inert mass would make a quite suitable substitute for a live test person. This should be investigated further because it is easy to "calibrate" a mass. Passive spring-mass dummies could extend the frequency range and could be calibrated in a traceable way [Howard and Stayner 1998], but difficulties have been found with controlling the damping elements. An active dummy can overcome this, but would be more difficult to calibrate in a process that allows traceability to primary standards. This aspect requires further work.

Conclusions

Gradual progress is being made toward reaching an acceptable reproducibility in testing suspended seats. Detailed guidance for test procedures is likely to help, but there are more steps to be made. Elimination of human test persons remains an outstanding goal.

Acknowledgment: The author wishes to thank CVG KAB Seating for supporting this work.

References

EN 30326-1 [1994]. Mechanical vibration – laboratory method for evaluating vehicle seat vibration (ISO 10326-1:1992). Part 1: Basic requirements.

Howard JC, Stayner RM [1998]. Dynamic dummies for seat testing: what the industry wants is a standard. In: 33rd U.K. Conference on Human Response to Vibration (Buxton, U.K.)

Mozaffarin A, Pankoke S, Bersiner F, Cullmann A [2007]. MEMOSIK V: Development and application of an active, three-dimensional dummy for measurement of vibration comfort on vehicle seats. In: Proceedings of the Third National Conference on Human Vibration (Dresden, Germany, October 8–9, 2007).

Nélisse H, Patra S, Boutin J, Boileau P-É, Rakheja S [2006]. Evaluation of two anthropo-dynamic manikins for seat testing under whole-body vibration. In: 41st U.K. Conference on Human Response to Vibration (Farnborough, U.K., September 20–22, 2006).

Suggs CW, Abrams CF, Stikeleather LF [1969]. Application of a damped spring-mass human vibration simulator in vibration testing of vehicle seats. Ergonomics *12*(1):79–93.

MULTIAXIS SEAT CUSHION TRANSMISSIBILITY CHARACTERISTICS

Suzanne D. Smith,[1] David R. Bowden,[2] and Jennifer G. Jurcsisn[1]

[1] Air Force Research Laboratory, Wright-Patterson Air Force Base, OH
[2] General Dynamics Advanced Information Systems, Dayton, OH

Introduction

The spectral characteristics of higher-frequency multiaxis military propeller aircraft vibration have been investigated [Smith et al. 2007]. The vibration properties of the coupled seat/cushion/human were further evaluated by applying a multiple-input/single-output model to estimate the system transfer matrix associated with exposure to multiaxis random vibration.

Methods

A U.S. Navy E-2C Hawkeye aircraft seat was tested with six cushion configurations and seven subjects in a multiaxis vibration facility. Cushion configurations A through E included five different seat pan cushions and the original seat back cushion with separate lumbar support. Cushion configuration F included the seat pan cushion used in E combined with a prototype seat back cushion. Triaxial accelerations were measured at the floor and on the left side of the seat near the base. Triaxial accelerometer pads measured the vibration between the human and cushions at the seat pan and seat back. Flat constant-bandwidth random vibration (1–80 Hz) was generated at 1.0 ms^{-2} rms to produce linearly independent combined-axis motions in the X, Y, and Z axes. The system transfer matrix is written as:

$$\begin{bmatrix} H_{xO} \\ H_{yO} \\ H_{zO} \end{bmatrix} = \begin{bmatrix} s_{xx} & s_{xy} & s_{xz} \\ s_{yx} & s_{yy} & s_{yz} \\ s_{zx} & s_{zy} & s_{zz} \end{bmatrix}^{-1} \begin{bmatrix} S_{xO} \\ S_{yO} \\ S_{zO} \end{bmatrix} \qquad (1)$$

where H_{xO}, H_{yO}, and H_{zO} represent the transfer functions (transmissibilities) between a single output, O, and inputs x, y, and z at the floor (ω not shown in Equation 1); s_{xx}, s_{xy},…s_{zz} are the auto- and cross-spectra between the input signals; and S_{xO}, S_{yO}, and S_{zO} are the cross-spectra between the three inputs at the floor and the output, O. The output, O, was measured in the X, Y, and Z directions at the seat pan, seat back, and seat base.

Results

Figure 1 illustrates the mean multiaxis transmissibilities for the major responses at the seat pan and seat back. All cushions tended to show a small amplification of the seat pan X/x, seat pan Z/z, and seat back X/x vibration in the vicinity of whole-body vertical resonance (4–8 Hz). The seat pan, seat back, and seat base showed a substantial peak in the Y/y transmissibility between 12 and 15 Hz. This peak was also observed in the seat back X/x transmissibility, but at a lower magnitude. All cushions showed increases in the seat base and seat pan X/x, Z/z, and X/z transmissibilities over a broad range between 30–50 Hz (Figure 1). This increase was also observed for the seat back Z/z. Higher seat pan X/x and seat pan X/z transmissibilities were observed at higher frequencies (>70 Hz). Cushion F showed significant damping in the higher-frequency seat back X/x and seat back X/z transmissibilities (Figure 1).

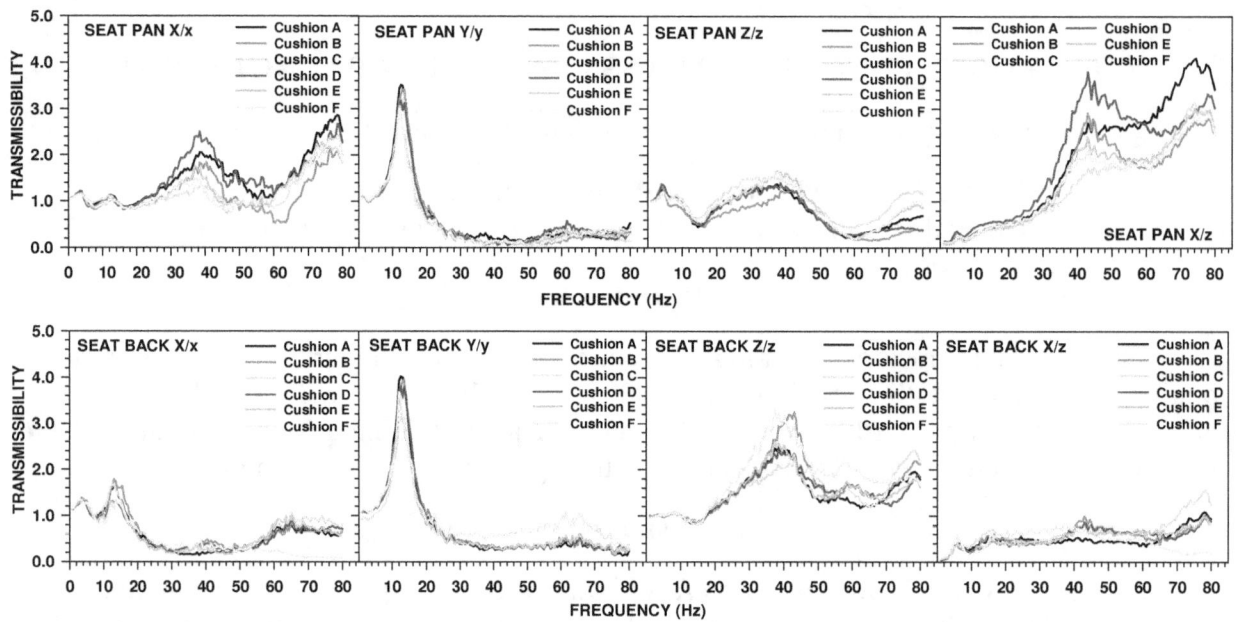

Figure 1.—Mean seat pan and seat back transmissibilities (seven subjects).

Discussion

The operational exposures [Smith et al. 2007] showed that the older original seat pan cushion (cushion A) produced significantly higher seat pan accelerations in the X direction at the blade passage frequency (BPF ~73.5 Hz) compared to the other tested cushions. This was predicted by the transmissibility data: cushion A produced significantly higher seat pan X/x transmissibility compared to cushions B and C (Figure 2) and produced significantly higher seat pan X/z transmissibility compared to all other cushions. The z-axis vibration may have the greatest influence on the higher-frequency X-axis seat pan vibration; the X/z transmissibilities were higher compared to X/x. However, the input vibration at the BPF was the highest in the X direction. The damping effect of the prototype seat back cushion (cushion F) was also observed at the BPF. The seat base transmissibility results did indicate that the E-2C seat structure was not rigid, particularly in the Y axis, and most likely influenced the behavior at the seat/occupant interfaces. The transmissibility data can be used to predict the multiaxis accelerations entering the human during operational exposures and target seat vibration mitigation strategies.

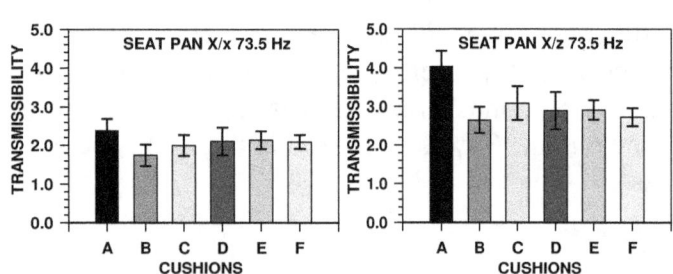

Figure 2.—Seat pan X/x and X/z transmissibility at 73.5 Hz.

Reference

Smith SD, Jurcsisn JG, Walker AY, Smith JA, Bowden DR [2007]. Dynamic characteristics and human perception of vibration aboard a military propeller aircraft. Wright-Patterson Air Force Base, OH: Air Force Research Laboratory, Human Effectiveness Directorate, Biosciences and Protection Division, Biomechanics Branch. Report No. AFRL-HE-WP-TR-2007-0114. NTIS No. ADA473700.

COMPARISON OF THE PERFORMANCE OF A PNEUMATIC ACTIVE SEAT SUSPENSION USING A DISCRETE AND FUZZY LOGIC-BASED CONTROL LAW

Bertrand Valero,[1] Farid Amirouche,[1] and Alan G. Mayton[2]

[1]Vehicle Technology Laboratory, University of Illinois at Chicago (UIC)
[2]National Institute for Occupational Safety and Health, Pittsburgh Research Laboratory, Pittsburgh, PA

Introduction

The torque generated by a heavy vehicle engine can lead to a high level of vibration inside the operator's cabin. It raises not only safety issues concerning the operator's ability to perform his or her task, but also health issues. Indeed, prolonged exposure to such vibration has been linked to an increase in the risk of lower back problems and other illnesses [Schwarze et al. 1998].

The addition of a suspension system between the operator's seat and the chassis of the vehicle lowers the level of vibration felt by the driver. Passive suspension designs have been introduced but have limitations. The skyhook control method [Karnopp et al. 1974] was used in this study, which investigated an active seat suspension under various conditions. The main idea behind the active suspension design is to introduce an actuator and adjust its damping properties depending on input signals from sensors placed on the seat and chassis of the vehicle [Valero et al. 2007].

The damping of the actuator can be set at different levels chosen from a predetermined set of configurations for the actuator [Valero et al. 2006]. The nonlinearity introduced by such an approach can lead to chattering due to the high-frequency switching in the actuator command. To achieve more linear behavior in the actuator, a control method using fuzzy logic concepts [Guglielmino et al. 2005] was implemented, with an active pneumatic suspension design developed at the UIC Vehicle Technology Laboratory.

Methods

The suspension system was tested at the National Institute for Occupational Safety and Health's Pittsburgh Research Laboratory. The suspension was bolted to an MTS shake table, and a mass of 40 kg was bolted to the top of the suspension system. The test protocol consisted of a sinusoidal vibration sweep of amplitude ±10 mm or ±20 mm and a frequency range of 1–8 Hz at 0.1-Hz intervals. The signals from two accelerometers placed on top of the suspension and onto the chassis were used as input by a control algorithm that returns a position for the valve using the fuzzy logic approach. Different variations of the control laws were tested by changing the number of membership functions for the fuzzy logic implementation. The pressure inside the actuator of the pneumatic suspension was also set to different levels. During a previous test session, the same testing protocol, suspension design, and conditions were used to evaluate the performance of the suspension. The only difference was the control method used; it included a discrete number of damping levels for the actuator. Subsequently, the efficiency of the fuzzy logic approach can be evaluated by comparing the results obtained during the two testing sessions.

Results

In both cases, when actuator pressure was increased, the results showed a reduction of the peak amplitude at the resonant frequency for all of the control strategies. The discrete approach showed a direct correlation between the resonance peak and the ability to select higher damping options using different valve positions. In the fuzzy logic approach, we observed better isolation at the resonant frequency when the number of membership functions considered in the control method was reduced. Of the two control approaches, the fuzzy logic approach showed better performance.

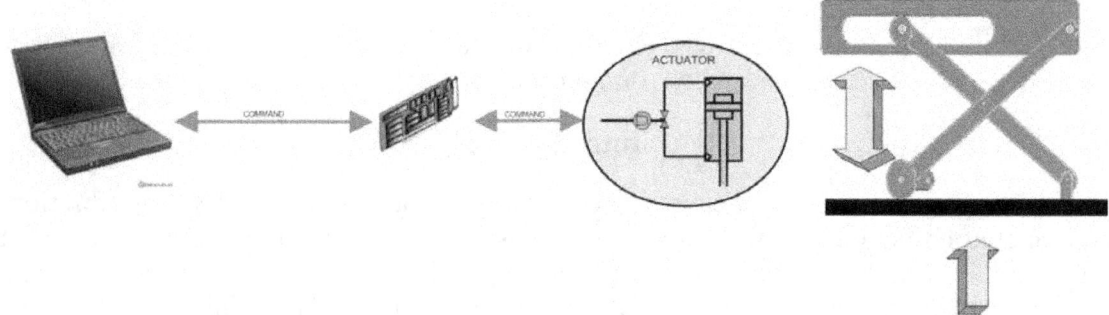

Figure 1.—Experimental setup.

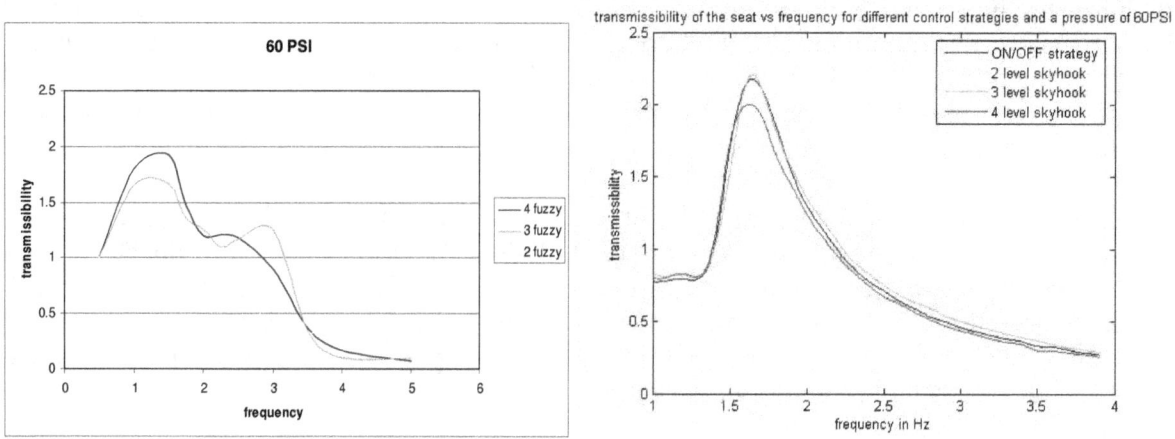

Figure 2.—Transmissibility of the seat with a fuzzy controller *(left)* and a discrete controller *(right)*.

References

Guglielmino E, Stammers CW, Stancioiu D, Sireteanu T, Ghigliazza R [2005]. Hybrid variable structure-fuzzy control of a magnetorheological damper for a seat suspension. Int J Vehicle Autonomous Syst *3*(1):34–46.

Karnopp DC, Crosby MJ, Harwood RA [1974]. Vibration control using semi-active force generators. ASME J Eng Ind *96*(2):619–626.

Schwarze S, Notbohm G, Dupuis H, Hartung E [1998]. Dose-response relationships between whole-body vibration and lumbar disk disease: a field study on 388 drivers of different vehicles. J Sound Vibration *215*(4):613–628.

Valero B, Amirouche FML, Mayton AG [2006]. Pneumatic active suspension design for heavy vehicle seats and operator ride comfort. In: Proceedings of the First American Conference on Human Vibration (June 5–7, 2006). Morgantown, WV: U.S. Department of Health and Human Services, Public Health Service, Centers for Disease Control and Prevention, National Institute for Occupational Safety and Health, DHHS (NIOSH) Publication No. 2006–140, pp. 38–39.

Valero B, Amirouche FML, Mayton AG, Jobes CC [2007]. Comparison of passive seat suspension with different configuration of seat pads and active seat suspension. Warrendale, PA: Society of Automotive Engineers, Inc., technical paper 2007-01-0350.

CHARACTERIZATION OF THE NOISE AND VIBRATION PRODUCED BY PORTABLE POWER TOOLS USED IN THE AUTOMOTIVE REPAIR INDUSTRY

Pierre Marcotte,[1] Rémy Oddo,[2] Jérôme Boutin,[1] and Hugues Nélisse[1]

[1]Institut de recherche Robert-Sauvé en santé et en sécurité du travail (IRSST), Montreal, Quebec, Canada
[2]Acoustics Group, University of Sherbrooke, Sherbrooke, Quebec, Canada

Introduction

Users of handheld power tools can be exposed to excessive levels of noise and hand-arm vibration. The current ISO standards used to evaluate the vibration and noise levels of these tools generally do not take into account the interaction between the tool and the worked piece. However, this interaction is important since, in some cases, the noise and vibration generated by these tools is strongly dependent upon the tool-worked piece interaction. Thus, the goal of this study was to develop new test benches that include the tool-worked piece interaction.

Methods

Test benches based on existing standards were used to measure the noise and vibration levels of three categories of handheld power tools currently used in auto repair shops and recognized as being potentially problematic in terms of noise exposure and hand-arm vibration. They consist of pneumatic wrenches, impact wrenches, and die grinders. They were tested according to the ISO 8662-7 and ISO 8662-13 standards to assess their vibration levels and to the ISO 15744 standard to assess their noise levels. ISO 8662-7 describes the operation of pneumatic and impact wrenches in a resistive setup at a constant rpm, while ISO 8662-13 describes the operation of die grinders with an artificial eccentric bit. In an additional step, the test benches were modified in order to include the interaction between the tool and the worked piece, thus better representing the operation of these tools in a real work environment. For the pneumatic wrenches and impact wrenches, the second-generation test bench consisted of screwing nuts on threaded rods, while for the die grinder, the second-generation test bench consisted of grinding, sanding, and cleaning a standardized piece of steel with a metal brush. Measurements of noise and hand-arm vibration upon various handheld power tools were also carried out in several garages in order to guide the development of these second-generation test benches.

Results

A summary of the noise and vibration levels obtained with the first- and second-generation test benches is shown in Table 1 for four different pneumatic wrenches. The vibration levels are higher with the first-generation test benches for all tested tools, while the sound power levels are similar between the two generation test benches. The results obtained with five different impact wrenches are shown in Table 2. For the impact wrenches, the vibration levels obtained with the two generation test benches are similar, while the sound power level is 4.6–7.2 dB(A) lower with the second-generation test benches. Table 3 shows a summary of the noise and vibration levels for four different die grinders using the first- and second-generation test benches. The levels reported for the second-generation test bench consist of the maximum values obtained from the three tasks (grinding, sanding, and cleaning). The results show similar sound power levels for all tested tools, but the vibration levels are much greater with the second-generation test bench.

Table 1.—Summary of laboratory measurements on pneumatic wrenches

Manufacturer	Model	Capacity (Nm)	Vibration (m/s^2, W_h)		Sound power (L_w, dB(A))	
			1st generation	2nd generation	1st generation	2nd generation
A	1	68	12.54	4.48	102.9	101.8
B	3	81	NA	5.67	NA	94.5
B	2	68	8.35	5.32	100.3	100.2
C	4	75	5.24	4.05	98.3	98.7

Table 2.—Summary of laboratory measurements on impact wrenches

Manufacturer	Model	Capacity (Nm)	Vibration (m/s^2, W_h)		Sound power (L_w, dB(A))	
			1st generation	2nd generation	1st generation	2nd generation
A	1	949	6.89	6.46	107.2	100.0
A	2	949	6.93	5.08	101.0	95.6
B	3	847	6.59	5.39	103.1	98.5
B	4	576	4.01	3.83	103.9	96.7
C	5	576	5.49	5.05	106.1	99.0

Table 3.—Summary of laboratory measurements on die grinders

Manufacturer	Model	Vibration (m/s^2, W_h)		Sound power (L_w, dB(A))	
		1st generation	2nd generation	1st generation	2nd generation
B	1	6.15	14.59	95.6	97.8
B	2	5.68	9.61	104.2	103.8
C	3	3.55	10.19	104.2	104.5
A	4	7.59	10.02	NA	100.3

Discussion

The results from the second-generation test benches show lower noise levels for impact wrenches, lower vibration levels for pneumatic wrenches, and higher vibration levels for die grinders. However, paired t-tests indicate that the differences in the vibration levels of the impact wrenches between the two generations are not statistically significant ($\alpha = 0.05$). Thus, new ISO test benches should include the interaction of the tool with the worked piece in assessing vibration levels of pneumatic wrenches and die grinders and noise levels of impact wrenches.

References

ISO [1997a]. Hand-held portable power tools – Measurement of vibrations at the handle – Part 7: Wrenches, screwdrivers and nut runners with impact, impulse or ratchet action. Geneva, Switzerland: International Organization for Standardization. ISO 8662-7:1997.

ISO [1997b]. Hand-held portable power tools – Measurement of vibrations at the handle – Part 13: Die grinders. Geneva, Switzerland: International Organization for Standardization. ISO 8662-13:1997.

ISO [2002]. Hand-held non-electric power tools – Noise measurement code – Engineering method (grade 2). Geneva, Switzerland: International Organization for Standardization. ISO 15744:2002.

Session III: Whole-Body Vibration I

Chair: Subhash Rakheja *
Co-Chair: N. Shibata

Presenter	Title	Page
A. G. Mayton National Institute for Occupational Safety and Health (NIOSH)	Assessment of Whole-Body Vibration Exposure on Haulage Trucks and Front-end Loaders at U.S. Aggregate Stone Operations	26
M. S. Contratto Caterpillar, Inc.	Evaluation of Vibration Exposure of Operators in Wheel Tractor-Scrapers	28
S. Pankoke Wölfel	Static and Dynamic Properties of Polyurethane Foams for Mobile Machine Static Seats and Their Effects on Operator Exposure to Whole-Body Vibration	30
H. W. Paschold Ohio University	Pilot Study of Whole-Body Vibration in Fully Automated Residential Solid Waste Collection	32
B. Bazrgari École Polytechnique	Biodynamics of the Human Trunk in Seated Whole-Body Vibration	34

* Author unable to attend conference.

ASSESSMENT OF WHOLE-BODY VIBRATION EXPOSURE ON HAULAGE TRUCKS AND FRONT-END LOADERS AT U.S. AGGREGATE STONE OPERATIONS

Alan G. Mayton,[1] Richard E. Miller,[2] and Christopher C. Jobes[1]

[1]National Institute for Occupational Safety and Health, Pittsburgh Research Laboratory, Pittsburgh, PA
[2]National Institute for Occupational Safety and Health, Spokane Research Laboratory, Spokane, WA

Introduction

Preliminary vibration data were obtained during a study conducted at two eastern quarries of a major U.S. aggregate stone producer. As a result of employee feedback concerning low back discomfort and bouncing/jarring, company management decided to evaluate seating and operator whole-body vibration (WBV) exposure at two quarries for drivers/operators on haulage trucks and front-end loaders. Two study hypotheses were presented: (1) older equipment at quarry 1 would show worse WBV exposure results than the newer haulage/loading machinery at quarry 2; (2) measurements of WBV exposure would support employee complaints of musculoskeletal symptoms, particularly for the low back, during haulage truck and front-end loader operation.

Quarry 2 operated with newer model equipment with greater hauling and loading capacity than quarry 1. All of the haulage trucks were rear-dump style, and the haulage trucks and their respective seats were in good working order. The travel routes and drivers were different between the two quarries. The roadways were dusty and required constant watering for dust abatement during data collection at quarry 1, whereas a steady rain caused wet conditions during 1 of 2 days of data collection at quarry 2. Data were collected on one pit front-end loader at quarry 1 and one pit and one plant/yard loader at quarry 2.

Methods

Vibration data were collected using PCB Piezotronics (Depew, NY) triaxial accelerometers (models 356B18, 356B40) with signal amplifiers (model 480E09) and filters (150-Hz low-pass, model 474M32) connected to a Sony digital data recorder (model PC208Ax, Sony Manufacturing Systems America, Lake Forest, CA). The mean sampling time for quarry 1 was 21 min (standard deviation (SD) = 13 min); mean sampling time for quarry 2 was 63 min (SD = 24 min). These were considered representative of exposures for the shift. WBV exposures for vehicle drivers/operators were assessed relative to ISO 2631-1 (1997) and the *European Union Good Practice Guide for WBV* [Griffin et al. 2006].

Results and Discussion

Table 1 describes the equipment operating at quarries 1 and 2 and shows the results from WBV exposure measurements collected for haulage trucks and front-end loaders. Weighted RMS accelerations (wRMS) and vibration dose values (VDVs) for the dominant axis of vibration and overall weighted total RMS accelerations (vector sums) were computed for comparison.

In five of six trials for quarry 1 haulage trucks, $wRMS_z$ exceeded the action level (0.5 m/s^2) recommended in the *European Union Good Practice Guide for WBV* [Griffin et al. 2006]. By contrast, in one of three trials for haulage trucks at quarry 2, $wRMS_z$ just barely exceeded this same action level. No haulage trucks at either quarry showed VDV_z that reached the 9.1 $m/s^{1.75}$ action level. Vector sums may indicate driver/operator discomfort or a "rougher" ride (ISO 2631-1 (1997)). For the older trucks at quarry 1, vector sums varied from 0.64 to 2.73 m/s^2 with a mean of 1.58 m/s^2 and 0.87 m/s^2 SD. Vector sums for quarry 2 haulage trucks exhibited less variation, ranging from 0.42 to 0.61 m/s^2 with a mean of 0.51 m/s^2 and 0.10 m/s^2 SD. In contrast to the quarry 1 pit front-end loader, the newer and larger-capacity pit loader at quarry 2 showed $wRMS_x$ at 1.21 m/s^2, exceeding both action and exposure levels specified in the *European Union Good Practice Guide for WBV* [Griffin et al. 2006]; VDV_x at 12.26 $m/s^{1.75}$ exceeded the action level. In comparing the mean $wRMS_z$ and vector sum accelerations in this study with Eger et al. [2006], quarry 2 haulage trucks were slightly higher, but the quarry 1 trucks were more than three to four times higher. In summary, the two hypotheses stated earlier seem true in view of the findings, although more data are needed to verify this.

Table 1.—Description of equipment with weighted RMS accelerations, vector sums, and vibration dose values (shading indicates dominant axis)

Mobile equipment	Manufacturer	Year	Capacity	$wRMS_x$ (m/s^2)	$wRMS_y$ (m/s^2)	$wRMS_z$ (m/s^2)	Vector sum (m/s^2)	Overall VDV ($m/s^{1.75}$)	VDV_x ($m/s^{1.75}$)	VDV_y ($m/s^{1.75}$)	VDV_z ($m/s^{1.75}$)
Quarry No. 1											
Haulage truck	A	1986	50.8 mt	0.48	0.56	1.00	1.15	5.21	1.97	2.06	4.36
				0.28	0.25	0.46	0.64	2.91	1.47	1.01	2.25
	A	1978	50.8 mt	0.73	0.74	1.83	2.73	8.28	3.07	3.31	6.94
				0.28	0.34	0.60	0.96	3.04	1.21	1.42	2.40
	A	1986	50.8 mt	0.45	0.48	0.98	1.40	4.86	1.80	1.80	4.15
				0.91	1.08	1.68	2.57	8.63	3.59	4.13	6.67
Front-end loader	B	1997	6.3 m^3	0.30	0.43	0.41	0.66	3.31	1.64	1.93	2.13
Quarry No. 2											
Haulage truck	C	2000	66.0 mt	0.25	0.27	0.51	0.49	8.98	3.26	3.25	7.71
	C	1999	66.0 mt	0.25	0.26	0.41	0.42	6.55	3.23	3.14	4.75
	B	2000	71.1 mt	0.30	0.37	0.43	0.61	7.27	3.33	4.17	4.94
Front-end loader	B	2005	10.7 m^3	1.21	1.02	0.75	1.24	17.58	12.26	10.30	7.27
	C	2000	5.4 m^3	0.32	0.28	0.36	0.36	6.19	3.80	2.97	3.87

References

Eger T, Salmoni A, Cann A, Jack R [2006]. Whole-body vibration exposure experienced by mining equipment operators. Occup Ergon 6(3/4):121–127.

Griffin MJ, Howarth HVC, Pitts PM, Fischer S, Kaulbars U, Donati PM, Bereton PF [2006]. Guide to good practice on whole-body vibration: non-binding guide to good practice with a view to implementation of directive 2002/44/EC on the minimum health and safety requirements regarding the exposure of workers to the risks arising from physical agents (vibrations). [http://www.humanvibration.com/EU/VIBGUIDE/WBV%20Good%20practice%20Guide%20v6.7g%20English%20070606.pdf]. Date accessed: May 2008.

ISO [1997]. Mechanical vibration and shock: evaluation of human exposure to whole-body vibration. Part 1: General requirements. Geneva, Switzerland: International Organization for Standardization. ISO 2631-1:1997.

EVALUATION OF VIBRATION EXPOSURE OF OPERATORS IN WHEEL TRACTOR-SCRAPERS

Michael S. Contratto and Jinyan (Christy) Du
Caterpillar, Inc., Peoria, IL

Introduction

Due to uncertain knowledge of human responses to vibration, both objective vibration measurements and subjective scaling are used to evaluate the vibration level to which machine operators are exposed. This study seeks to develop a model to describe the operator's perceived discomfort as a function of the measured acceleration so that vibration comfort level can be predicted by the acceleration measurements. This was done by (1) identifying the vibration measurement that seems to be the best correlation with subjective response, (2) investigating effects caused by different experimental design variables on subjective response and vibration acceleration, and (3) using linear regression to model the relationship between the selected vibration measurement and the subjective response.

Methods

Nine operators were selected to run a wheel tractor-scraper passing over a bump. Four variables were investigated, including two bump heights, two load conditions (empty and full), left or right tire crossing over the bump, and five speed levels. Acceleration was measured at the seat base and seat pad using triaxis accelerometers with a sampling rate of 500 Hz. In order to identify which frequency weighting and analysis method gave the best correlation with subjective rating, four frequency-weighted filters were applied on the raw measurement: ISO, ISOM (modification to ISO 2631 weight curves), jerk, and 12 Hz (Figure 1).

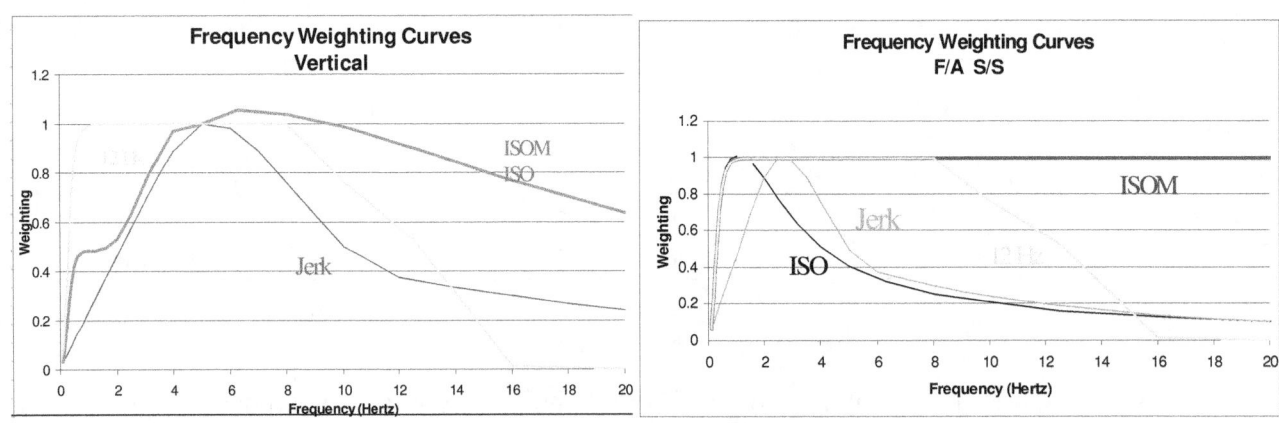

Figure 1.—Frequency weighting curves. (F/A = fore/aft; S/S = side/side.)

For each filter, the minimum, maximum, range, root-mean-square (RMS), vibration dose value (VDV), and maximum transient (x,y,z), maximum of the RMS value of three vectors (x,y,z), and the vector sum RMS were calculated. A 1–9 rating scale was used to determine the operators' subjective assessment of vibration comfort.

Results

Statistic analysis was performed using Minitab. Analyses of Pearson's correlation test showed measurements from the seat pad had higher correlation with the subjective measurements than those from the seat base. The correlation coefficients with all four frequency-weighted filters were similar. The highest correlation coefficients were around 0.68–0.7 with vector sum RMS, maximum RMS (x,y,z), VDV, and maximum transient vibration value (MTVV). Jerk data showed slightly higher correlation than others. Also, smaller deviations were shown in the vector sum RMS and maximum RMS (x,y,z) than in the VDV and MTVV; therefore, the jerk vector sum RMS data were selected for further statistical analysis.

Analysis of the main effects demonstrated that speed, bump height, and the tire passing over the bump have significant effects on both acceleration measurements and subjective response (P-value < 0.001). A regression model based on the group average of each test run showed a significant relationship between the jerk vector sum RMS and operator subjective rating, while the regression model of the individual data showed large variation between the vibration measurements and the subjective rating ($R^2 = 48\%$) (Figure 2).

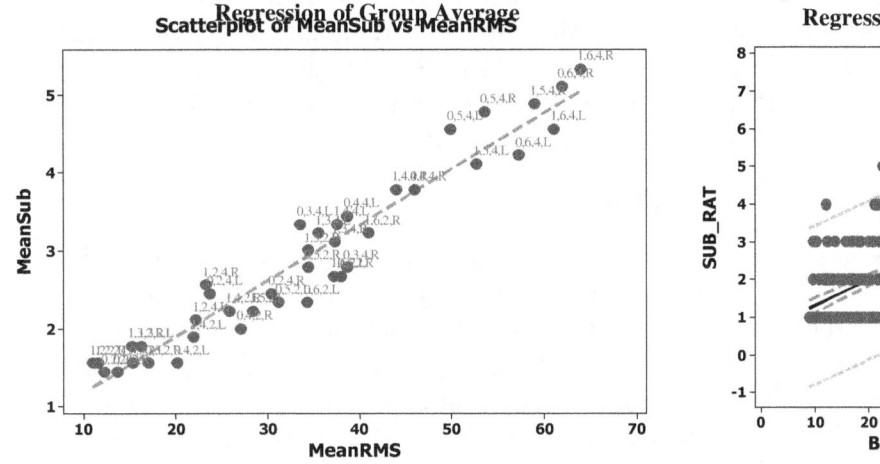

Figure 2.—Regression of group average and individual measurements.

Normalizing the subjective rating data helped increase the correlation with the acceleration measurement (r = 0.785). The regression model using the normalized subjective data was also better than using the nonnormalized data ($R^2 = 62\%$) (Figure 3). However, due to the high operator-to-operator variation even with normalized data, the upper boundary of the data set was used as the linear regression function to predict the comfort level using the acceleration measurements.

Discussion

This study concluded that the difference between frequency weightings is not significant. Because the variance of the subjective response to comfort level is large, the correlation between the subjective response and acceleration measurement is not very good. The variance also affects the accuracy of the regression model. Normalization does help decrease variation of the subjective scaling data, but improvement is not significant.

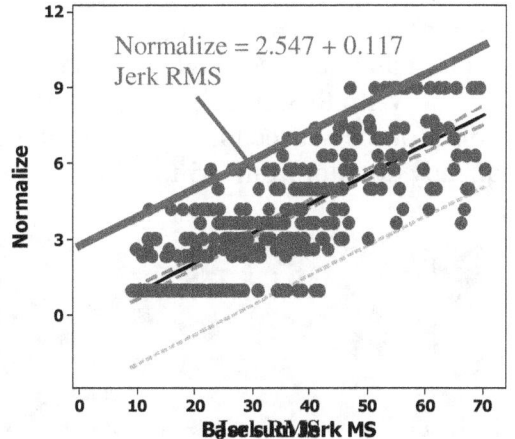

Figure 3.—Regression model of normalized rating and the RMS value.

STATIC AND DYNAMIC PROPERTIES OF POLYURETHANE FOAMS FOR MOBILE MACHINE STATIC SEATS AND THEIR EFFECTS ON OPERATOR EXPOSURE TO WHOLE-BODY VIBRATION

S. Pankoke,[1] M. S. Contratto,[2] A. Striegel,[2] and J. Hofmann[1]

[1]Wölfel Beratende Ingenieure GmbH + Co. KG, Höchberg, Germany
[2]Caterpillar, Inc., Peoria, IL

Introduction

Seat cushions have a dramatic impact on both the ride experienced by the operator and the operator's subjective opinion of the machine ride. Static seats for mobile machines and commercial vehicles, i.e., seats that do not have an explicit spring and damper suspension, may provide isolation of the operator from vibrations only by the stiffness and damping properties of the polyurethane foams. It is desirable to analyze the effect of foam properties on vibration isolation behavior when used in a static seat with a human.

Methods

Polyurethane foams are investigated by static and dynamic compression tests of quad-shaped foam samples. A hydraulic test rig compresses the samples between two rigid plates (Figure 1). Compression (mm) and compression force (N) are measured and evaluated. From the raw data, static hysteresis plots and plots of the frequency-dependent complex Young's Modulus are computed. The data are then used for the identification of material models for a dynamic finite-element simulation using Abaqus and the dynamic occupant model CASIMIR in configuration m50 (1.75-m body height, 75-kg body mass) [Pankoke and Siefert 2006].

With such identified material models for the polyurethane foams, a simplified model of the occupied seat is set up. The seat is modeled with a quad-shaped foam cushion and backrest; the seat and back are considered to be rigid. The dynamic occupant model, CASIMIR configuration m50, represents the operator (Figure 2). The sitting posture of CASIMIR is adapted to typical seat and backrest angles of operator seats.

With this model, a static seating simulation is performed to find the appropriate operating point for the succeeding dynamic simulation. The dynamic simulation is carried out in the frequency domain from 0.5 to 15 Hz, using vertical (z) excitation at the cushion base.

Figure 1.—Hydraulic test rig with foam.

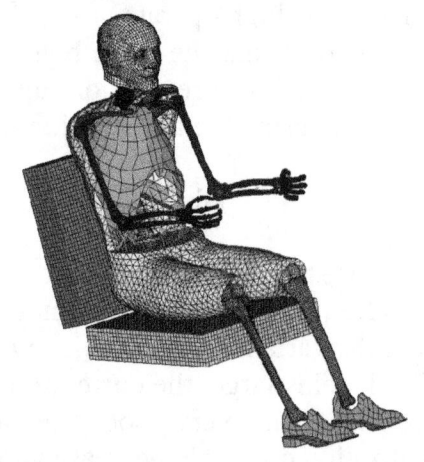

Figure 2.—Simplified seat with CASIMIR.

Results

Results from the static material tests show the typical S-shaped hysteresis curves of polyurethane foams (Figure 3). Differences occur between loading and unloading, building the hysteresis. For the static material model, the average between loading and unloading is used as identification target. Identification results are very close to this target.

The dynamic material model is based on the complex Young's modulus, which depends on the polyurethane material (Figure 4), static precompression, frequency, and the dynamic vibration magnitude.

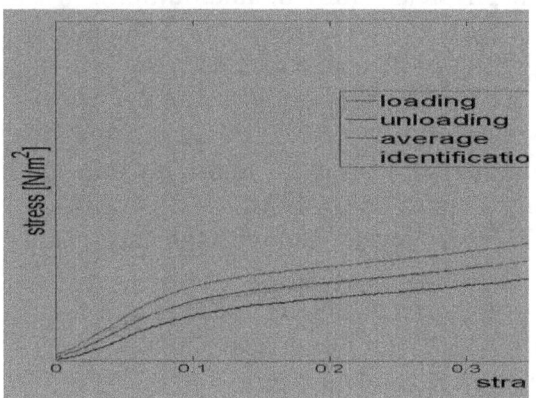

Figure 3.—Static hysteresis with identification.

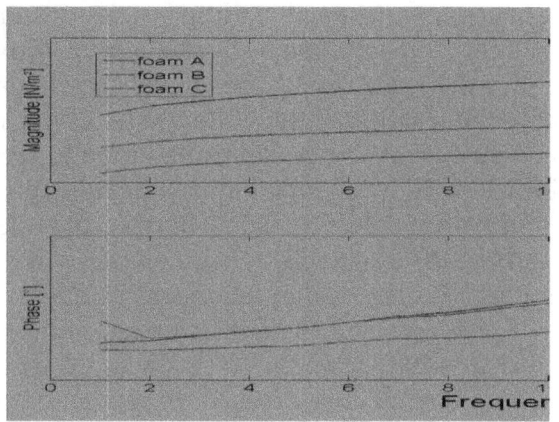

Figure 4.—Complex Young's modulus.

Differences in dynamic foam characteristics result in differences in the isolation behavior of the foams as expressed by the vertical seat transfer function $H_S(s)$, defined as the complex ratio between the vertical vibration response $Q(s)$ on the cushion surface and the vertical vibration excitation $Q^0(s)$ at the seat base ($s = i\Omega$ is the Laplace variable): $H_S(s) = Q(s)/Q^0(s)$. Figure 5 shows the differences in $H_S(s)$.

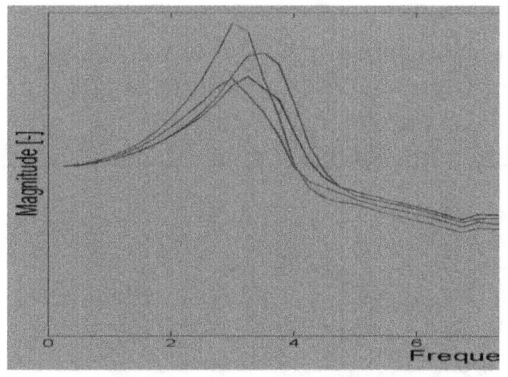

Figure 5.—Seat transfer functions $H_S(s)$ for different foams.

Discussion

With the above results for $H_S(s)$ for different foams, it is now possible to identify the best cushion property to minimize operator vibrations. It is foreseeable that this will depend on the machine type, since different machines have different typical excitation spectra at the seat base. The identification of the best foam for different machine types (wheel loaders, tractor scrapers, track-type tractors) will be the next step in this study.

Reference

Pankoke S, Siefert A [2006]. Simulation of human motion, muscle forces, and lumbar spine stresses due to whole-body vibration: application of the dynamic human model CASIMIR for the development of commercial vehicles and passenger cars. In: Proceedings of the First American Conference on Human Vibration (June 5–7, 2006). Morgantown, WV: U.S. Department of Health and Human Services, Public Health Service, Centers for Disease Control and Prevention, National Institute for Occupational Safety and Health, DHHS (NIOSH) Publication No. 2006–140, pp. 46–47.

PILOT STUDY OF WHOLE-BODY VIBRATION IN FULLY AUTOMATED RESIDENTIAL SOLID WASTE COLLECTION

Helmut W. Paschold
School of Health Sciences, Ohio University, Athens, OH

Introduction

Residential solid waste collection methods range from manual to semiautomated to automated loading. Manual loading is accomplished by a crew physically moving and dumping waste containers into hoppers of trucks. Semiautomated systems require manual movement and placement of the waste containers at the truck hopper so a hydraulic or mechanical device can lift and empty the waste containers. In a fully automated system, a driver collects solid waste from containers using a joystick-controlled robotic arm, without leaving the vehicle cab. Full automation eliminates safety and health hazards resulting from manual lifting, walking, and direct contact with the solid waste, leading solid waste collection management to believe that most job risks have been eliminated. However, two new hazard exposures were introduced with the automated system: joystick manipulation and whole-body vibration (WBV).

Several studies have evaluated solid waste collection trucks and WBV. Maeda and Morioka [1998] reported vibration exposure exceeding ISO 2631 recommended limits in the mode of manual collection. With semiautomated collection, Melo and Carvalhinho [2006] reported WBV doses exceeding the action value in 5 of 13 garbage trucks, while Bovenzi et al. [2006] found WBV levels below the action value in 6 trucks.

In the automated method, the driver positions the truck in approximate alignment with the container and extends a robotic arm. This arm grasps, lifts, and dumps the filled container into a hopper at the top of the truck (see Figure 1). Automated operators are exposed to the vibrations from the robotic arm lifter. Also, it seems that the automated vehicles tend to have more stops and starts as the driver aligns with each individual container.

Methods

During the summer of 2007, vibration measurements were made in a total of four vehicles during a 2-day period. WBV measurements were obtained with Quest Technologies (QT) HAVPro human vibration meters with QT seat pad triaxial accelerometers. The data buffers recorded a series of 24 consecutive 20-min sets, equivalent to a full workday in a residential urban neighborhood on relatively flat, well-paved streets.

Figure 1.—Automated collection truck *(left)* with a closeup of the robotic arm *(right)*.

Results and Discussion

The data in Table 1 and Figure 2 are representative of other WBV values for collection activity obtained in this pilot study. Lower levels were found during travel, idle time, and landfill transport. The higher values of this study in the horizontal (x and y) axes, when compared with those previously reported for solid waste truck operation, may be explained by the robotic arm activity or increased vehicle starts and stops. Additional data analysis is being performed to evaluate the combined daily data set for an accurate daily exposure value. This pilot study identifies a need to obtain improved operation mode detail from the operator or an assigned observer. Additional variables to consider in future studies include the effects of collection route characteristics (terrain, roads, home spacing), equipment type, robotic arm operation, and seat transmission.

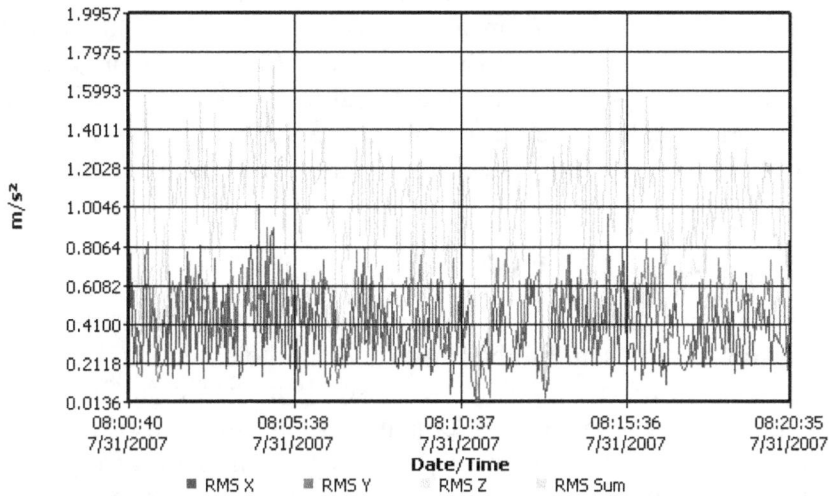

Figure 2.—Results of one 20-min period during residential solid waste collection.

Table 1.—Tabular results from Figure 2 (m/s^2)

	X	Y	Z	Sum
Amin	0.0166	0.0151	0.0879	—
Amax	1.0095	0.8986	0.8734	—
A(8)	0.4120	0.5123	0.5542	1.0744

References

Bovenzi M, Rui F, Negro C, D'Agostin F, Angotzi G, Bianchi S, Bramanti L, Festa G, Gatti S, Pinto I, Rondina L, Stacchini N [2006]. An epidemiological study of low back pain in professional drivers. J Sound Vibration *298*(3):514–539.

Maeda S, Morioka M [1998]. Measurement of whole-body vibration exposure from garbage trucks. J Sound Vibration *215*(4):959–964.

Melo RB, Carvalhinho F [2006]. Whole-body vibration exposure of waste truck drivers. In: Proceedings of the 16th World Congress of the International Ergonomics Association (Maastricht, Netherlands, July 10–14, 2006). Elsevier.

BIODYNAMICS OF THE HUMAN TRUNK IN SEATED WHOLE-BODY VIBRATION

B. Bazrgari,[1] A. Shirazi-Adl,[1] and M. Kasra[2]

[1]Department of Mechanical Engineering, École Polytechnique, Montreal, Quebec, Canada
[2]Department of Mechanical Engineering, McMaster University, Hamilton, Ontario, Canada

Introduction

Long-term exposure to whole-body vibration (WBV) may increase risk of spine disorders [Lings and Leboeuf-Yde 2000], an association that still remains controversial. To elucidate the likely causative role of WBV in low-back disorders, however, one needs accurate knowledge on muscle forces, spinal loads, and trunk stability. Previous trunk WBV model studies have either neglected or oversimplified the trunk redundancy, and trunk stability has not been addressed. In continuation of our earlier isometric/transient investigations using a kinematics-driven approach [Arjmand and Shirazi-Adl 2006; Bazrgari et al. 2007], this work aims to evaluate spinal loads, muscle forces, and stability in a seated subject under a vehicular random vertical base excitation with ~ ±1 g peak acceleration contents [Seidel et al. 1997]. The effects of posture and coactivity were studied. It is hypothesized that high acceleration contents (shocks) increase spinal loads and risk of injury.

Methods

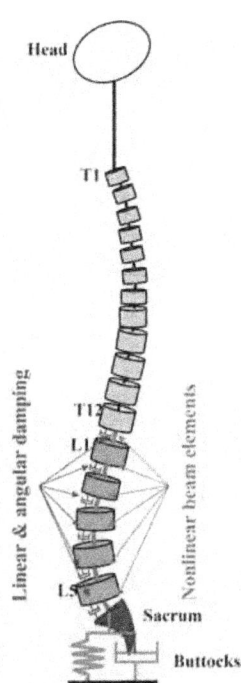

A three-dimensional head-pelvis model (Figure 1) made of six non-linear deformable beams to represent lumbar discs and rigid elements to represent the rest of the head-thorax segments was used [Bazrgari et al. 2007]. The buttocks were modeled by a connector element (compression only) with nonlinear properties. Masses, mass moments of inertia, and dampers were introduced at different levels based on the literature. A sagittally symmetric muscle architecture with 56 muscle fascicles was used. Trunk kinematics, distributed gravity, and base excitation were applied onto the model. The nonlinear response, muscle forces, and spinal loads were computed using the kinematics-driven approach, which satisfies equilibrium at all levels/directions while accounting for time-varying muscle forces. Stability analyses were performed taking a linear force-stiffness relationship for muscles with a critical stiffness coefficient (q) computed using nonlinear and linear (perturbation and eigenvalue) analyses. The input excitation was taken from measurements collected at the vehicle seat-driver interface (Figure 2) [Seidel et al. 1997]. Two lumbar postures of erect and flexed were considered; in the latter, the lordosis was flattened by 10°. The effect of antagonistic activity in abdominal muscles and changes in buttocks stiffness on response was also investigated.

Figure 1.—Head-pelvis model.

Results

Base net moments remained positive (flexion) throughout vibration and were larger in the flexed posture (Figure 2). Spinal compression/shear forces reached their maximum values at the lowermost L5-S1 level in the flexed posture. The critical (minimum) muscle stiffness coefficient to maintain stability was substantially influenced by muscle activity, posture, and coactivity in the abdominal muscles. The flexed posture improved trunk stability. Introduction of antagonistic abdominal activity substantially increased the trunk stability margin in the erect posture by diminishing critical q from 89 to 21 at the onset of vibration and from 131 to 39 at the time of peak upward acceleration. Reductions in buttocks stiffness decreased peak accelerations, natural frequency, muscle forces, and spinal loads.

Discussion

The computed vertical accelerations at the L3 and head (Figure 2) were in good agreement with measurements [Seidel et al. 1997]. Incorporation of the buttocks with realistic properties diminished the first vertical natural frequency from ~12 to 5.5 Hz, in agreement with the literature. The spinal compression/shear forces at the L5-S1 increased significantly because of inertial forces—from the respective static values of 535 and 222 N in the erect posture and 717 and 297 N in the flexed posture to peak values of 1,131 and 508 N in the former and 1,608 and 771 N in the latter posture. Based on the premise that excessive loads on the spine could cause injury and considering the mild shock contents as compared with values of ±2–6 g reported in off-road and industrial vehicles, WBV can be considered as a risk factor. The 10° flattening in the lumbar lordosis increased the net moment, muscle forces, and passive spinal loads while substantially improving trunk stability. Similarly, the introduction of antagonistic coactivity in the abdominal muscles increased extensor muscles activity, spinal loads, and the stability margin. An increase in trunk stability occurs at the cost of higher passive loads and hence greater risk of tissue injury, suggesting a tradeoff between opposing demands.

Acknowledgment: Supported by NSERC-Canada and the Aga Khan Foundation.

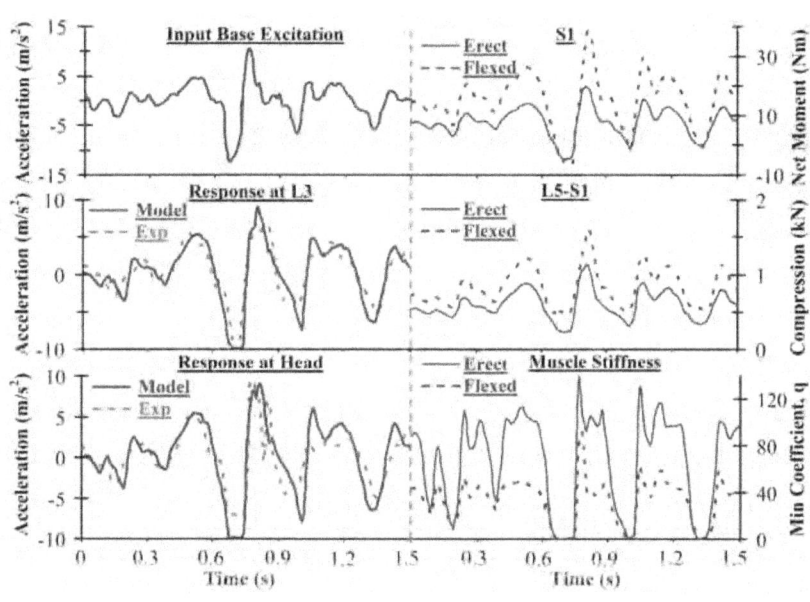

Figure 2.—Input base excitation and acceleration (left) with predictions for for postures (right).

References

Arjmand N, Shirazi-Adl A [2006]. Model and in vivo studies on human trunk load partitioning and stability in isometric forward flexions. J Biomech 39(3):510–521.

Bazrgari B, Shirazi-Adl A, Arjmand N [2007] Analysis of squat and stoop dynamic liftings: muscle forces and internal spinal loads. Eur Spine J 16(5):687–699.

Lings S, Leboeuf-Yde C [2000]. Whole-body vibration and low back pain: a systematic, critical review of the epidemiological literature 1992–1999. Int Arch Occup Environ Health 73(5):290–297.

Seidel H, Blüthner R, Hinz B, Schust M [1997]. Stresses in the lumbar spine due to whole-body vibration containing shocks (final report). Dortmund/Berlin, Germany: Schriftenreihe der Bundesanstalt für Arbeitsschutz und Arbeitsmedizin.

Session IV: Human Body Modeling and Vibration I

Chair: Suzanne D. Smith
Co-Chair: H. P. Wölfel

Presenter	Title	Page
A. Mozaffarin Wölfel	MEMOSIK V: Development and Application of an Active Three-Dimensional Dummy for Measuring Vibration Comfort on Vehicle Seats	37
B. Bazrgari École Polytechnique	Spinal Loads, Muscle Forces, and Stability under Axial Shocks	39
H. P. Wölfel Darmstadt University of Technology	Application of the Response Spectrum Analysis Method for the Evaluation of Human Response to Vibration	41
W. J. Pielemeier Ford Motor Co.	Hybrid Modal Analysis of Occupied Seats	43
M. A. Książek	Models of a Human Operator as a Push Force Regulator in a Human-Handtool System	45

MEMOSIK V: DEVELOPMENT AND APPLICATION OF AN ACTIVE THREE-DIMENSIONAL DUMMY FOR MEASURING VIBRATION COMFORT ON VEHICLE SEATS

Amin Mozaffarin and Steffen Pankoke
Wölfel Beratende Ingenieure GmbH + Co. KG, Höchberg, Germany

Introduction

The unavoidable occurrence of vehicle vibrations affects operating safety, health, and comfort in both commercial vehicles and passenger cars. In addition to the vibration exposure at the interface between the seat and the occupant, seat transfer functions are usually measured for the rating of these effects. Because the transfer functions depend on the dynamic loading of the seat, the application and validity of such measurements are only suggestive when using human test subjects. However, the use of human subjects is complicated by various restrictions and disadvantages. The development of a vibration dummy is presented here in order to address the difficulties in the experimental rating of whole-body vibrations.

Active Vibration Dummy MEMOSIK V

The newly developed MEMOSIK V (Figure 1) is an actively controlled, three-dimensional vibration dummy that reproduces the vibration behavior of a seated occupant for the x-, y-, and z-direction by its dynamic masses. It is able to replace the occupant for measurements in a vehicle as well as on test rigs. The adoption of the dynamic vibration behavior to the three standard mass percentiles (f05 – 5th female; m50 – 50th male; and m95 – 95th male), as well as to a particular individual, is easily made through software modifications to the controller setup. The appropriate static adoption masses are added or removed.

Figure 1.—MEMOSIK V vibration dummy.

The postural variability is accommodated by the complete design range of vehicle seats from SUV to roadster. MEMOSIK V comes up with an integrated measurement system and enables the repeatable measurement of vibration exposure and seat transfer functions in all three spatial directions x, y, and z.

Active Model Approach

The active model approach of MEMOSIK V is based on the enhancement of an initially passive, ground-excited single degree of freedom (SDOF) oscillator by an actuator and an appropriate control system in order to design a modifiable frequency response (Figure 2).

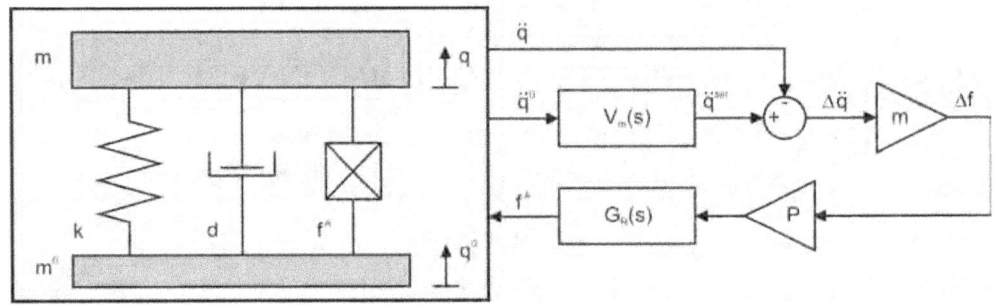

Figure 2.—Equivalent model of the active ground-excited SDOF.

The dynamic mass $M(s)$ of this structure results from the contributions of the ground mass m^0 and the moving mass m, weighted with the amplification function $V_m(s)$. The amplification function $V_m(s)$ represents the relation between the response displacement $Q(s)$ of the moving mass and the excitation displacement $Q^0(s)$ of the ground mass ($s = i\Omega$ is the Laplace variable).

$$M(s) = m^0 + mV_m(s) \qquad (1)$$

$$V_m(s) = Q(s)/Q^0(s) \qquad (2)$$

In conjunction with the controller, the actuator enables the modification of the amplification function and, thus, the dynamic mass can be adjusted to a given objective function (Figure 3).

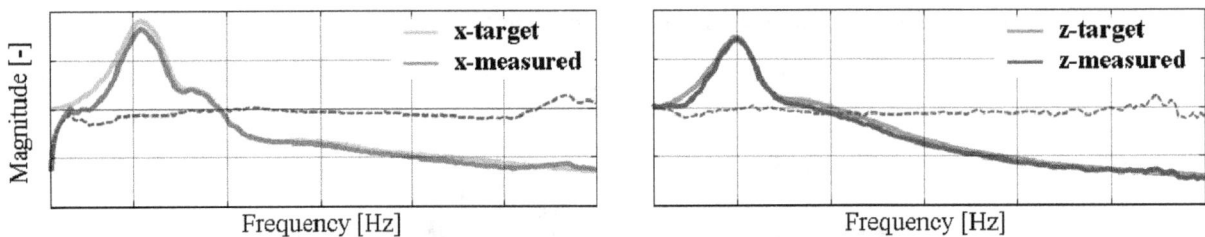

Figure 3.—Amplification functions $V_m(s)$ for x and z, target and measurement.

Modal Identification

As a dummy development base, a measuring program was arranged in order to acquire the dynamic masses (x, y, and z) of human subjects seated on a rigid test seat. Subsequently, the required objective functions were deduced using a modal system identification algorithm to parameterize the measurement results for the implementation in the control system. The modal identification is based on the approximation of the measured data by the sum of multiple SDOF oscillators. Each single oscillator represents a specific mode of the system.

Bibliography

Cullmann A [2002]. An active vibration dummy of the seated occupant (in German). Düsseldorf, Germany: VDI report No. 492.

Patent Specification DE 19858517 C1 [1998]. Active vibration model of the human, especially of the seated occupant (in German).

SPINAL LOADS, MUSCLE FORCES, AND STABILITY UNDER AXIAL SHOCKS

B. Bazrgari and A. Shirazi-Adl

Department of Mechanical Engineering, École Polytechnique, Montreal, Quebec, Canada

Introduction

Effective management of back disorders depends in part on accurate estimation of spinal loads, muscle forces, and trunk stability in various occupational conditions. Existing biomechanical models of spine have either neglected or oversimplified the nonlinear passive resistance, muscle forces, complex geometry/loading/dynamics of the spine, and equilibrium in all spinal levels and directions. Our group has introduced an iterative kinematics-based finite-element approach to address the existing shortcomings in spinal models. Thus far, the model has successfully been applied to simulate a number of isometric and dynamic manual material-handling tasks [Arjmand and Shirazi-Adl 2005; Bazrgari et al. 2008]. The current work presents a novel application of the kinematics-based approach to investigate the seated trunk response (e.g., spinal loads, muscle forces, and trunk stability) to vehicular vibration with axial shock contents (~4 g) at different frequencies. It is hypothesized that the magnitude and frequency of the base excitation, via muscle recruitments (a parameter generally overlooked in earlier models), markedly affect the response and risk of injury.

Methods

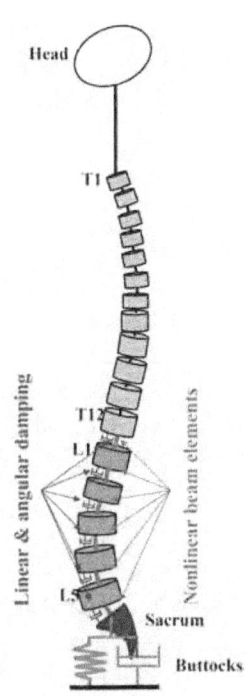

Figure 1.—Finite-element model.

The kinematics-driven approach combines experimental and model studies to alleviate the kinetic redundancy of the trunk musculoskeletal system. Measured trunk kinematics (e.g., rotations at different vertebral levels), along with external/gravity loads, were prescribed into a nonlinear, finite-element, three-dimensional head-pelvis model (Figure 1) made of nonlinear deformable beams to represent lumbar segments and rigid elements for head-thorax segments [Bazrgari et al. 2008]. The buttocks were modeled by a connector element (compression only) with nonlinear properties. Masses, mass moments of inertia, and dampers were introduced at different levels based on the literature. The model was based on data from different sources and close to the 85th percentile in the 40–50 years age group. A sagittally symmetric muscle architecture with 56 muscle fascicles was used. Temporal variation of response is calculated using the nonlinear iterative kinematics-driven approach. Trunk stability was subsequently investigated taking a linear force-stiffness relationship for muscles with a critical stiffness coefficient (q) computed using nonlinear and linear (perturbation and eigenvalue) analyses. Axial base excitations with ~4 g peaks and frequencies of ~4 and 20 Hz were considered based on data at a vertical sinusoidal shaker table [Robinson 1999]. The effect of antagonistic activity in abdominal muscles was also investigated.

Results

Spinal loads were much greater for the excitation frequency of 4 Hz compared to 20 Hz. Maximum net moment at the S1 was also greater in the former (69 versus 37 Nm). Accordingly, compression/shear forces at the L5-S1 disc (i.e., the level with maximum loads) for base excitations with 4- and 20-Hz frequencies were, respectively, 3,468/1,236 N and 1,933/697 N.

Contribution of muscle forces to these loads was significant and generally exceeded that of inertial loads (Figure 2). The critical (minimum) muscle stiffness coefficient reached values of 350 and 150 at peak acceleration periods, respectively, under 4- and 20-Hz frequencies, indicating risk of instability. In contrast, the system was rather stable in periods with negative acceleration or active abdominal muscles. Introduction of antagonistic abdominal coactivity in the case under 4-Hz frequency substantially increased the trunk stability margin by diminishing critical q from 88 to 12 at the onset of vibration and from 350 to 63 at the time of peak upward acceleration. Amelioration in spinal stability due to antagonistic muscle coactivities was, however, at the expense of increases in spinal loads.

Discussion

Under various activities such as those involving vehicular vibrations, the spinal loads substantially alter depending on the temporal variation of activation in the trunk muscles. The relative contributions of gravity, inertial, and muscle forces to the peak compression/shear forces at the L5-S1 for the excitation frequency of 4 Hz were, respectively, 10/11%, 39/48%, and 51/41%, which altered to 9/10%, 35/44%, and 56/46% in the presence of abdominal coactivity and to 18/20%, 32/40%, and 50/40% as the excitation frequency increased to 20 Hz (Figure 2). Large shock contents (~4 g in the current study) in off-road and military vehicles can significantly increase spinal loads and, hence, risk of injury especially when applied at frequencies in the neighborhood of trunk natural frequency (~5 Hz). Such loading conditions also substantially deteriorate system stability, thereby demanding compensatory antagonistic muscle coactivity that further increases spinal loads and risk of spinal injuries.

Acknowledgment: Supported by NSERC-Canada and the Aga Khan Foundation.

References

Arjmand N, Shirazi-Adl A [2005]. Biomechanics of changes in lumbar posture in static lifting. Spine *30*(23):2637–2648.

Bazrgari B, Shirazi-Adl A, Trottier M, Mathieu P [2008]. Computation of trunk equilibrium and stability in free flexion-extension movements at different velocities. J Biomech *41*(2): 412–421.

Robinson DG [1999]. The dynamic response of the seated to mechanical shock (Dissertation). Burnaby, British Columbia, Canada: Simon Fraser University.

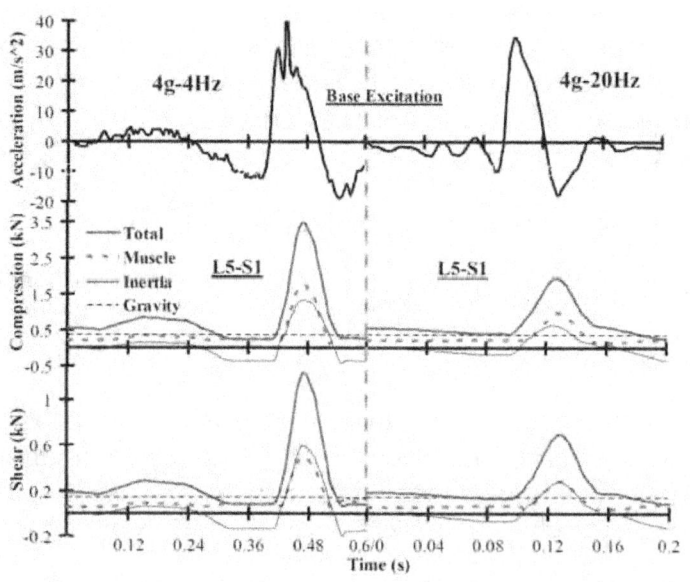

Figure 2.—Input and predictions for excitation contents of 4 *g*-4 Hz *(left)* and 4 *g*-20 Hz *(right)*.

APPLICATION OF THE RESPONSE SPECTRUM ANALYSIS METHOD FOR THE EVALUATION OF HUMAN RESPONSE TO VIBRATION

H. P. Wölfel[1] and J. Hofmann[2]

[1]Darmstadt University of Technology, Darmstadt, Germany
[2]Wölfel Beratende Ingenieure GmbH + Co. KG, Höchberg, Germany

Introduction

In view of the large number of workers with back pain, a reliable preventive evaluation of vibration exposures on humans at work is of great importance. In addition, the number of international regulations and guidelines dealing with the protection of the worker against vibration exposures that have been decreed during previous years make it obvious that we need to understand the relationship between occupational exposures and back pain.

Current evaluations of workplaces with regard to their potential for health impairment are based on methods introduced in ISO 2631-1 (1997), which often lead to questionable results [Seidel et al., in press]. The main reason originates from the attempt to reduce the vibration exposure to a single numerical value (root-mean-square (rms), root-mean-quad, maximum transient vibration value, vibration dose value), which facilitates the classification, but at the expense of reliability. However, reliability of the classification is critical when dealing with health-related matters.

Methods

This paper introduces the response spectrum analysis method, which offers a safer and more reliable basis for evaluating vibration exposures with regard to their potential for health impairment. The response spectrum (RS) should not be confused with a Fourier spectrum, since the RS characterizes the *response* of a vibratory system and not the excitation. To compute the RS, a set of single degree of freedom systems (SDOFSs) with different natural frequencies f_n is excited by the exposure. By taking the maximum value of the response of each SDOFS and plotting it over the natural frequency f_n, one obtains a curve with the maximum responses, the response spectrum (see Figure 1). Since the maximum response of the human directly correlates to the potential for health impairment, the response spectrum directly shows whether or not a signal can cause health impairment. Since the *maximum* value is selected and it is not considered *how often* this value appears within the signal, the RS can be used for evaluating the *direct risk* of impairment due to single exposure. A dose consideration is thus not possible. Such a dose consideration can be done by picking each local maximum A_i and minimum separated by a zero-pass and performing a higher power summation (see Equation 1). Such operation results in the response dose spectrum (RDS). The RDS differs from the RS since it considers how often an event of certain intensity of human response appears. The RDS D_n for the SDOFS with natural frequency f_n depends on the power of summation m: a higher power m puts emphasis on single events with high amplitude, while a lower power m stresses events with lower amplitude, which appear more frequently.

$$D_n = \left[\sum_i \left(A_{i\,n} \right)^m \right]^{\frac{1}{m}} \qquad (1)$$

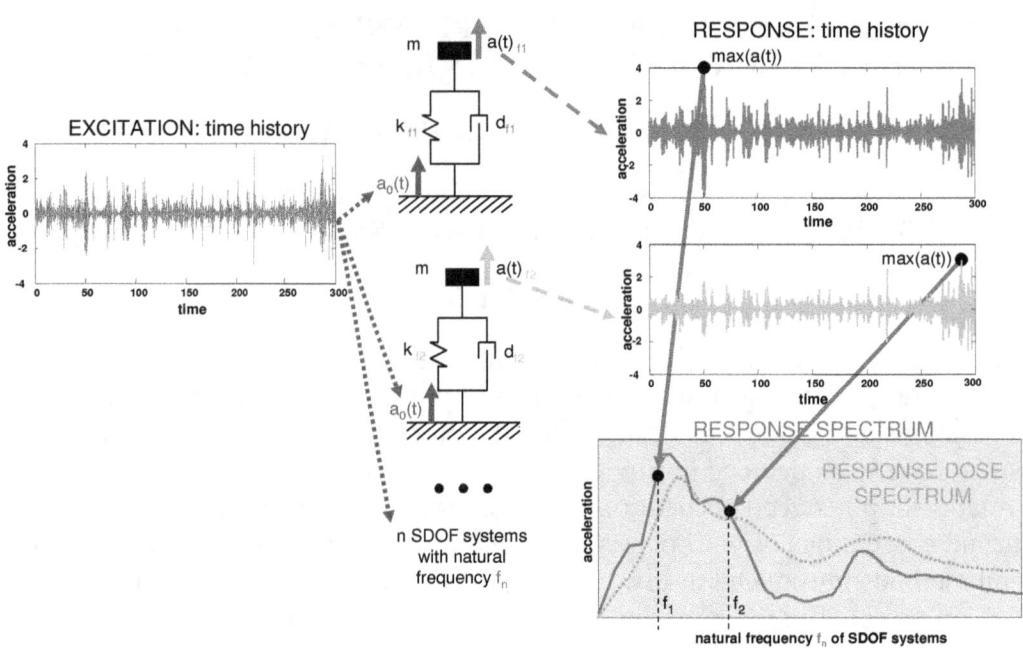

Figure 1.—Calculation of the response spectrum and the response dose spectrum [Gupta 1990].

Discussion

By performing numerous calculations with selected exposure signals, it can be shown that the response spectrum analysis method produces much more differentiated results than the single value-based methods. The main disadvantage of those methods, which is the disability to differ between signals that provide identical rms value (for example) but do have significant differences in their spectrum and thus in their response of the exposed human, can be avoided using RS and RDS for evaluation.

With definition of a maximum curve not to be exceeded, this method not only allows for evaluation of risk of health impairment, but is also suitable for standardization as well. A similar method considering only the RS and not the RDS is standardized in earthquake engineering.

Several industrial applications for the RS and RDS method are apparent. Due to the ability to consider the response over the whole spectrum at one glance, it is easy to figure out in which direction further developments should tend. A seat manufacturer, for example, can decide in which frequency ranges the seat behavior must be enhanced to reduce vibration effects on the operator. Single-value methods do not provide such information. Such application requires definition of standard excitations (typical for the vehicle under investigation) and appropriate limit curves.

References

Gupta AK [1990]. Response spectrum method in seismic analysis and design of structures. Boston, MA: Blackwell Scientific Publications, Inc.

ISO [1997]. Mechanical vibration and shock: evaluation of human exposure to whole-body vibration. Part 1: General requirements. Geneva, Switzerland: International Organization for Standardization. ISO 2631-1:1997.

Seidel H, Hinz B, Hofmann J, Menzel G [in press]. Intraspinal forces and health risk caused by whole-body vibration: predictions for European drivers and different field conditions. Int J Ind Ergon.

HYBRID MODAL ANALYSIS OF OCCUPIED SEATS

William J. Pielemeier
Ford Motor Co., Dearborn, MI

Introduction and Background

The measurement of the transmission of vibration through occupied vehicle seats has proven useful for understanding the effect of the seat on both vibration comfort and health effects [Griffin 1990]. Occupied seat transmissibility provides an input/output relationship between the attachment points to the floorpan and the interface between the seat and occupant. However, it provides no information on how the mechanical systems in the seat contribute to the transmissibility. Modal analysis is a standard tool for analyzing the linear vibration properties of mechanical systems. However, traditional modal analysis performed on a seat under the conditions of an occupied seat transmissibility test (i.e., mounted to a shaker system and occupied by a human) presents problems. Two new methods are described here that solve those problems and produce high-quality results.

A normal modal analysis excites the system with a wideband force input such as a modal hammer, or a shaker and stinger. With multiple inputs, all must be independent and uncorrelated for the system to be solved. Output accelerations are measured and transfer functions computed from the input forces to output accelerations. These functions are solved for the system modal properties, particularly mode frequencies, shapes, masses, and damping. Applying normal modal analysis to analyze a seat in the context of an occupied seat transmissibility test presents problems. With the seat attached to a rigid shaker system at four points, it is not mathematically possible for the force inputs to all four attachment points to be independent and uncorrelated.

Operational modal analysis is an alternate method that is often applied when test conditions do not allow control or measurement of input forces, such analysis of buildings, flying airplanes, etc. It uses only output accelerations, specifying some as references. Cross spectra are computed between the references and other outputs, forming a system of equations that can be solved for the mode frequencies, shapes, and damping, but not modal masses.

In practice, operational modal analysis results can suffer from the fact that the spectral properties of the excitations are not removed from cross spectra as they are when transfer functions are computed in a normal modal analysis. The presence of the input spectra in the cross spectra acts as noise when analyzing for the system properties in the output. When applied to modal analysis of a seat transmissibility test, this compromises the results of the analysis due to poor signal-to-noise ratio, because many system modes are highly damped and the input spectra fluctuations act as lightly damped peaks, which are fitted first by the modal analysis.

Methods

A pair of new methods of occupied seat modal analysis take advantage of the fact that the seat behaves as an approximately linear system when excitation dynamic range is limited. If an occupied seat is excited in pure translation (i.e., one axis at a time), then all of the attachment inputs receive the same acceleration. If all of the outputs are divided by a unitless copy of the acceleration input, linearity gives the outputs that will be obtained if a perfectly flat unity magnitude white-noise acceleration input was employed. If cross spectra are computed treating the input acceleration as an output reference, this process removes the fluctuations of the input spectrum and results in a high-quality operational modal analysis. A related idea can be applied to normal modal analysis as follows. An acceleration input can be substituted for the force input so that the standard transfer function computation removes the input spectral fluctuations and the normal modal equations can still be solved, but lack modal masses as with operational modal analysis. With the single-axis input condition described above, this acceleration input can be treated as a single input (even though at four points), so that dependent equations are not obtained. This is called hybrid modal analysis because it is a hybrid of normal and operational methods.

Results

Figure 1 shows the green line fit of five modes to the red vertical output at the seat cushion using a standard operational modal analysis. The vertical axis is in mg^2 for the cross spectrum between the vertical output and the wideband input used as a reference for operational modal analysis. Note the many sharp peaks due to the superposition of the excitation spectrum on the system modes, which the analysis fits with multiple lightly damped modes. Figure 2 is the same seat and excitation, analyzed with the second method described above, a normal modal analysis with a single vertical acceleration input. The vertical axis is the unitless transfer function amplitude ratio. Five modes are again used to fit the acceleration/acceleration transfer function obtained. Note the removal of the excitation artifacts and the superior fit of the magnitude and phase, with the analysis now fitting the more highly damped system modes. Fifty-four output points distributed over the entire seat along all three axes, including transmissibility outputs at the cushion and back of the seat system, gave similar improvements.

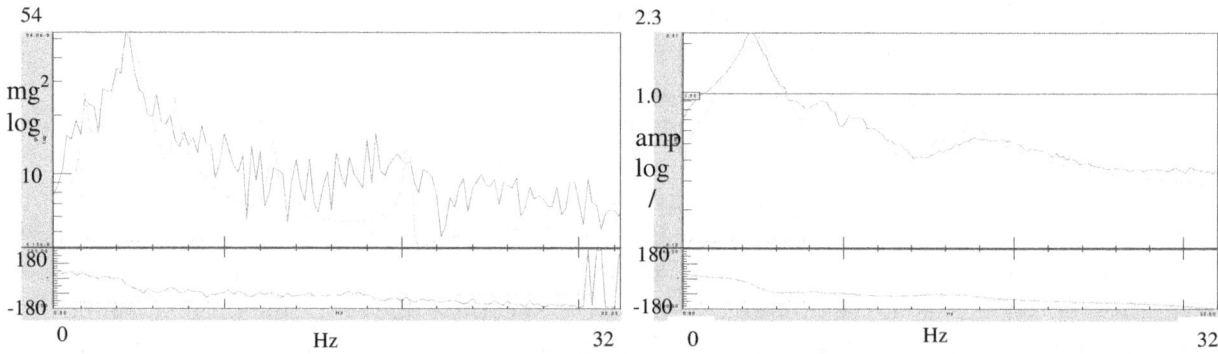

Figure 1.—Vertical cushion output (red) amplitude *(above)* and phase *(below)* fitted with five modes of an operational modal analysis (green).

Figure 2.—Same vertical cushion and excitation as in Figure 1, but treated with hybrid modal analysis.

Discussion

Animation of the mode shapes from the modal analysis identifies the mechanical sources of the resonances in the transmissibility functions. The results above clearly indicate that these alternatives to traditional methods of modal analysis provide an excellent fit to experimental data and the desired link to the mechanical behavior of the system.

Reference

Griffin MJ [1990]. Handbook of human vibration. London: Academic Press, pp. 387–413.

MODELS OF A HUMAN OPERATOR AS A PUSH FORCE REGULATOR IN A HUMAN-HANDTOOL SYSTEM

Marek A. Książek, Daniel Ziemiański, Zygmunt Basista, Janusz Tarnowski
Krakow University of Technology, Krakow, Poland

Introduction

The objective of our work was the synthesis of a model of human operator eye-hand reactions for assumed *a priori* working conditions with handtools. The data obtained from experimental investigations of a human operator's reactions were analyzed and modeled. The experiments were conducted on a stand specially designed for 14 volunteers. The registered output signals were approximated by functions that allowed the estimation of their corresponding transfer functions.

Experimental Investigations

During the experiment, a standing human operator applied measured and registered push force on a horizontal handle. The force was the result of screen observation where reference step and real push forces were shown side by side.

The signal of the step input force with a value of 100 N appeared unexpectedly on the screen. It was tracked by the operator, whose task was to realize, in the same period of time, a similar pressure force on the handle. The system measuring the force realized by the operator consisted of strain gauge force sensors. The output analog signal was proportional to the measured value of force. The exemplary time history of input and output signals, chosen from the set of analyzed signals, with marked characteristic points and parameters, is shown in Figure 1.

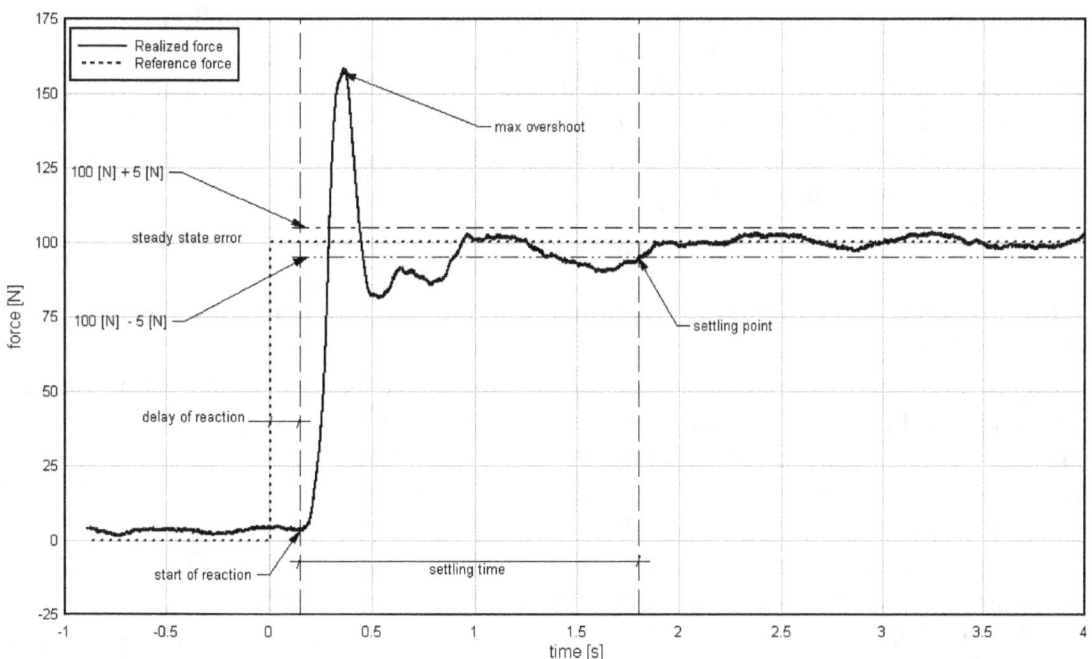

Figure 1.—Exemplary time history of step reference and forces realized by human operator.

The following parameters were assessed and calculated for each output signal: time delay of operator's reaction, rise time, peak amplitude, overshoot, steady-state error, and settling time.

Method of Modeling of Human Reactions

A diagram of the control system modeling human reactions is shown in Figure 2. The synthesis of the transfer function K(s) of the system was done by minimization of the integral quadratic performance index of error e(t) with an assumption of its physical realizability. The error e(t) denotes the difference between the real, registered, and approximated output signals.

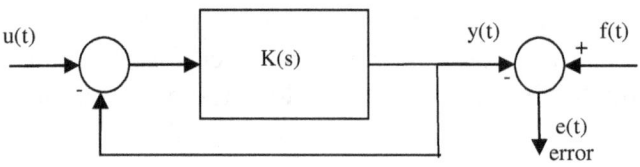

Figure 2.—Diagram of the investigated system.

Discussion of Results

An inspection analysis of the registered output signals shows that, in reality, each output signal should be modeled by a different transfer function. Among the analyzed signals, some different types can be specified. In the present approach, different Laplace transformations were applied to process different signals resulting from different types of human reactions. Such models allow more precise formulation of human reactions and criteria with regard to many aspects of work comfort with handtools. Mathematical modeling of the tool and human operator is indispensable for analyzing the dynamic and ergonomic aspects of the handtool-operator system as a whole. It has been shown that such modeling allows for better formulation of work comfort criteria and final estimation of design criteria of "human-friendly" handtools. First, the problems of reducing harmful vibration and vibration isolation must be addressed. Future experimental measurements will allow for the construction of a human operator model suitable for dynamic analysis of a human-machine system.

Bibliography

Alexík M [2000]. Modelling and identification of eye-hand dynamics. Simulat Pract Theory 8(1–2):25–38.

Basista Z [2006]. Simulation investigation of an active vibration protection system of an operator of a hand-held percussive tool. Eng Trans (Polish Acad Sci) 54(3):173–187.

Basista Z, Książck M [2004]. Estimation of comfort parameters of an active vibration isolation system of handle of percussive power tool. In: Proceedings of INTER-NOISE 2004 (Prague, Czech Republic, August 22–25, 2004).

de Vlugt E, Schouten AC, van der Helm FCT [2002]. Adaptation of reflexive feedback during arm posture to different environments. Biol Cybern 87(1):10–26.

McRuer D [1980]. Human dynamics in man-machine systems. Automatica 16(3):237–253.

This paper is supported by Polish Scientific Committee: PB 1255/T02/2007/32.

Session V: Whole-Body Vibration II

Chair: Bernard Martin
Co-Chair: Neil Mansfield

Presenter	**Title**	**Page**
N. Shibata Japan National Institute of Occupational Safety and Health	ISO 2631-1 Based Ride Comfort Evaluation of a Wheelchair Secured on Day Care Vehicles	48
A. Øvrum University Hospital of North Norway	Whole-Body Exposure from Heavy Loading Vehicles with Different Risk Assessment Outcomes Using Recommendations in the ISO-2631 and ISO-8041 Standards	50
L. F. Silva * Federal University of Itajubá	Study on Combined Exposure to Noise and Whole-Body Vibration and Its Effects on Workers' Hearing	52
N. Mansfield Loughborough University	The Compensatory Control Model as a Framework for Understanding Whole-Body Vibration	54

* Author unable to attend conference.

ISO 2631-1 BASED RIDE COMFORT EVALUATION OF A WHEELCHAIR SECURED ON DAY CARE VEHICLES

N. Shibata and S. Maeda

Japan National Institute of Occupational Safety and Health, Kawasaki, Japan

Introduction

Wheelchairs are an indispensable means of transport for those who have difficulty in walking and driving without the aid of mechanical equipment. Nowadays, a great number of people receive day care services, which include transport of older and/or disabled people by car while seated in wheelchairs secured on the vehicles. Recently, wheelchair users have complained of ride discomfort during transport by day care vehicles, which includes a fear of losing balance, swaying, falling forward, etc. Vibration transmitted through some surfaces in contact with the human body (e.g., seat back, seat surface, floor, and footrest) may cause such complaints. In this study, 12-axes vibration measurements based on ISO 2631-1 (1997) were performed to examine the effects of vibration on ride comfort for two types of day care vehicles.

Methods

The experiments were performed with one healthy female subject (aged 29 years). The subject had no prior experience of exposure to high levels or long periods of whole-body vibration occupationally or during leisure time activities. The subject was asked to sit in the wheelchair attached to the vehicle floor while driving these vehicles. The subject used a seat belt to secure her body in the wheelchair seat.

Figure 1 shows a 12-axes vibratory acceleration measurement system set up on the manual wheelchair used in this study. Two triaxial seat transducers were secured to the wheelchair: one on the seat back, the other on the supporting feet surface. Also, a six degree of freedom seat pad accelerometer was secured on the seat surface of the wheelchair. For each transducer, measurement directions were carefully adjusted to the basicentric axes of the human body in seated position specified in ISO 2631-1. Measurements of acceleration were obtained for vibration along the 12 axes with a sampling frequency of 1024 Hz.

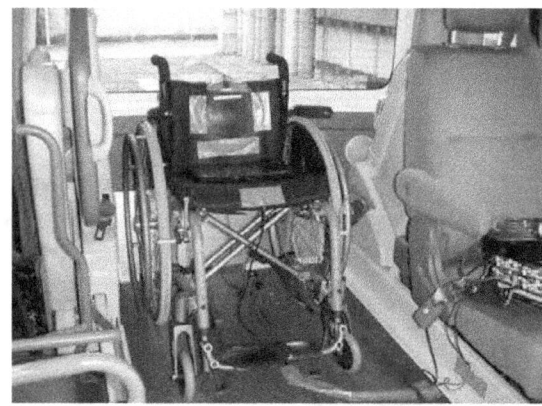

Figure 1.—Setup of the 12-axes vibratory acceleration measurement system.

Two types of vehicles, a large (vehicle 1) and a small one (vehicle 2), typically used in day care services in Japan, which can carry one wheelchair with the user seated in the wheelchair, were evaluated in this study. These day care vehicles were driven on two types of roads: alleys and trunk roads (highways).

Results and Discussion

As shown in Figure 2, vehicle 1 showed much higher total vibration values on alleys than trunk roads. By contrast, vehicle 2 did not show a significant difference in total vibration values between alleys and trunk roads. Also, total vibration values observed in vehicle 1 were higher than those observed in vehicle 2 for both trunk roads and alleys. This can be explained mainly by a difference in the mechanical structure of these two vehicles: vehicle 1 had a higher center of gravity position than vehicle 2. This structural characteristic caused vehicle 1 to vibrate more roughly and have larger displacements than vehicle 2, thereby adversely affecting the comfort of those riding in this vehicle.

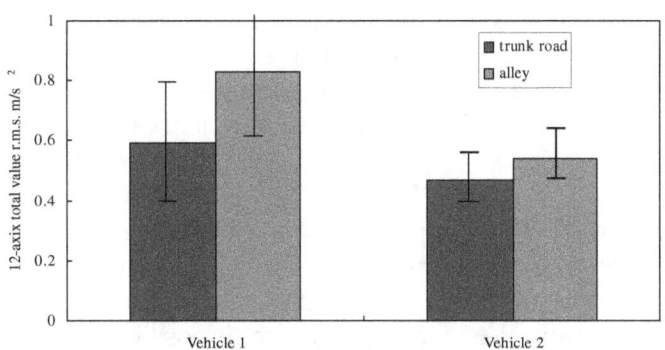

Figure 2.—Total vibration values obtained for two types of day care vehicles.

Regardless of vehicle type and road type, the frequency-weighted acceleration value in the x direction at seat back was the largest among the 12 acceleration components. In rotation effects, the acceleration value in pitching was the highest among the three rotation components. These acceleration components are likely to be high when the gradient of the road changes, as well as when vehicles start and stop. Reports have indicated that a number of wheelchair users felt extremely uncomfortable and sometimes in danger of falling forward. Indeed, some wheelchair users have fallen forward while riding in day care vehicles and have sustained minor injuries. In this regard, our results support the facts.

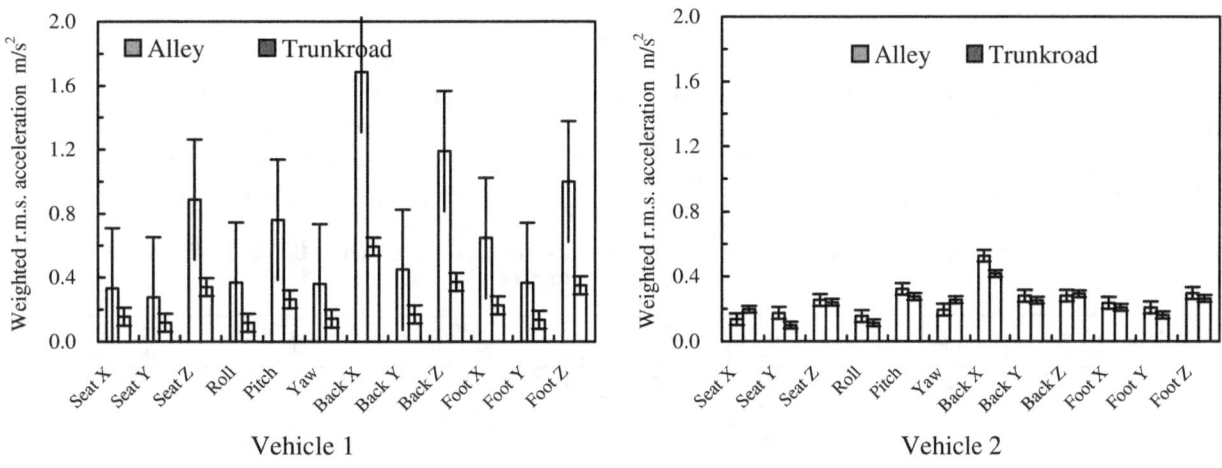

Figure 3.—Frequency-weighted rms acceleration for each axis.

Reference

ISO [1997]. Mechanical vibration and shock: evaluation of human exposure to whole-body vibration. Part 1: General requirements. Geneva, Switzerland: International Organization for Standardization. ISO 2631-1:1997.

WHOLE-BODY EXPOSURE FROM HEAVY LOADING VEHICLES WITH DIFFERENT RISK ASSESSMENT OUTCOMES USING RECOMMENDATIONS IN THE ISO-2631 AND ISO-8041 STANDARDS

A. Øvrum,[1] M. Skandfer,[1] A. Nikanov,[2] L. Talykova,[2] S. Syurin,[2] T. Khokhlov,[2] A. Vaktskjold[3]

[1]Department of Occupational and Environmental Medicine, University Hospital of North Norway, Tromsø, Norway
[2]Kola Research Laboratory for Occupational Health, Kirovsk, Russia
[3]Nordic School of Public Health, Gothenburg, Sweden

Introduction

Whole-body vibration (WBV) is a phenomenon to which vehicle drivers in the mining industry are exposed on a daily basis. The purpose of this project was to assess WBV exposure in a nonsimulated setting, applying portable equipment during real work cycles in underground and open mines on the Kola Peninsula. We calculated different possible risk assessments applying the ISO standard 2631-1 for load-haul-dump (LHD) vehicles. The results represent part of a cooperative project between the Department of Occupational and Environmental Medicine at the University Hospital of North Norway and the Kola Research Laboratory for Occupational Health in Kirovsk, Russia.

Methods

WBV exposure was measured in the OAO Apatity underground mine on the two most commonly used types of LHDs: TORO 007D, TORO 400E, one large transport truck; TORO 40D and K14 transport locomotives, operating in cycles performing loading, driving, dumping, and returning. The ground qualities ranged from soft to rocky surface. A total of 13 measurements were made. Measurement periods ranged from 10 to 52 min, with one to three cycles per period. The measurement, processing, analysis, and exposure assessment methods followed the guidelines of ISO 2631-1 (1997) and ISO 8041 (2005).

Results

Results from the measurements are shown in Table 1. To calculate A(8) (time-weighted acceleration over 8 hr), a time factor of T = 5 hr was used. The ISO standard 2631-1 allows for results with these characteristics to be presented in three ways: A(8) root-mean-square (rms) m/s^2, A(8) vibration dose value (VDV) m/s$^{-1.75}$, and vector sum (x, y, and z) A(8) rms (m/s^2).

Table 1.—Results from the WBV exposure measurements on the TORO 007D, TORO 400E, TORO 40D, and K14

LHD vehicle	Minutes	Cycles	A(8)x rms	A(8)x VDV	A(8)y rms	A(8)y VDV	A(8)z rms	A(8)z VDV	Vector sum A(8) rms
TORO 007D	10	3	0.56	4.12	0.44	3.43	0.76	7.80	1.23
TORO 007D	24	3	0.72	6.68	0.57	6.12	0.73	7.98	1.45
TORO 007D	17	3	0.81	6.68	0.47	3.95	0.98	10.27	1.61
TORO 400E	11	3	0.88	6.92	0.51	4.68	1	11.85	1.71
TORO 400E	11	3	0.72	5.45	0.47	3.63	0.52	4.36	1.30
TORO 400E	16	3	0.90	7.98	0.64	5.30	0.72	9.95	1.67
TORO 400E	24	3	0.82	7.81	0.65	6.08	0.83	9.64	1.65
TORO 40D	36	1	0.53	5.61	0.53	5.76	1.12	18.72	1.52
TORO 40D	34	1	0.72	7.45	0.57	5.69	1.05	14.93	1.64
TORO 40D	27	1	0.45	4.79	0.49	5.10	0.89	11.61	1.26
K14	32	1	0.20	3.01	0.12	1.83	0.19	4.20	0.37
K14	27	1	0.26	3.87	0.14	1.64	0.31	4.53	0.51
K14	52	1	0.2	4.05	0.1	2.39	0.29	12.8	0.42

Discussion

The number, types of vehicles, cycles, and days measured are considered adequate and representative in order to make a qualified risk assessment. The recommended basic risk evaluation method in ISO 2631-1 is used to present the results in Table 1 as A(8) rms, with results comparable to findings in other studies [Boileau et al. 2006; Shrawan 2004]. The characteristics of data for LHDs can also be presented using A(8) rms vector sum if two or more axes are comparable (ISO 2631-1, NOTE in 7.2.2), as seen in Table 1, or using A(8) VDV (if the crest factor is high, occasional shocks, transient vibration) (ISO 2631-1, 6.3, NOTE in 6.2.2) [Griffin 2004]. Depending on which method is applied, the LHD values provide different risk evaluations related to limit and action values (Table 2). Presented as dominant axis A(8) rms, the LHD values are below limit value, but above action value (0.5 m/s^2). Presented as vector sum A(8) rms, the LHD values are above limit value (1.15 m/s^2). Presented as A(8) VDV, the data are above and below the action value. Our WBV data on LHDs have substantial vibration levels in the x and z directions. The assessment method should therefore rely on the vector sum rather than the dominant axis, reducing the number of risk assessment methods for LHDs to one.

Table 2.—A(8) rms risk evaluation of TORO 007D, TORO 400E, TORO 40D, and K14 and risk evaluation of TORO 007D, TORO 400E using other methods in ISO 2631-1

Risk evaluation		Below action value	Above action value	Above exposure limit value
	A(8) rms	0 — 0.5 m/s^2	0.5 m/s^2 — 1.15 m/s^2	1.15 m/s^2 — 2.0 m/s^2
A(8) rms Highest axes	TORO 007D TORO 400E TORO 40D K14	$x_{0.2}$ $x_{0.29}$ $x_{0.31}$	$x_{0.73}$ $x_{0.76}$ $x_{0.98}$ $x_{0.72}$ $x_{0.83}$ $x_{0.90}$ $x_{1.0}$ $x_{0.89}$ $x_{1.05}$ $x_{1.12}$	
A(8) rms Vector sum	TORO 007D TORO 400E TORO 40D K14		$x_{0.37}$ $x_{0.42}$ $x_{0.51}$	$x_{1.23}$ $x_{1.45}$ $x_{1.61}$ $x_{1.3}$ $x_{1.65}$ $x_{1.67}$ $x_{1.74}$ $x_{1.26}$ $x_{1.52}$ $x_{1.64}$
	A(8) VDV	0 — 9.1 m/s$^{-1.75}$	9.1 — 21 m/s$^{-1.75}$	21 — 30 m/s$^{-1.75}$
A(8) VDV Highest axes	TORO 007D TORO 400E TORO 40D K14	$x_{7.8}$ $x_{7.98}$ $x_{5.45}$ $x_{7.81}$ $x_{7.98}$ $x_{4.2}$ $x_{4.9}$	$x_{10.27}$ $x_{11.85}$ $x_{11.61}$ $x_{14.93}$ $x_{18.72}$ $x_{12.8}$	

References

Boileau P-É, Boutin J, Eger T, Smets M [2006]. Vibration spectral class characterization of long haul dump mining vehicles and seat performance evaluation. In: Proceedings of the First American Conference on Human Vibration (June 5–7, 2006). Morgantown, WV: U.S. Department of Health and Human Services, Public Health Service, Centers for Disease Control and Prevention, National Institute for Occupational Safety and Health, DHHS (NIOSH) Publication No. 2006–140, pp. 14–15.

Griffin MJ [2004]. Minimum health and safety requirements for workers exposed to hand-transmitted vibration and whole-body vibration in the European Union: a review. Occup Environ Med *61*:387–397.

Shrawan K [2004]. Vibration in operating heavy haul trucks in overburden mining. Appl Ergon *35*(6):509–520.

STUDY ON COMBINED EXPOSURE TO NOISE AND WHOLE-BODY VIBRATION AND ITS EFFECTS ON WORKERS' HEARING

Luiz Felipe Silva[1] and René Mendes[2]

[1]Federal University of Itajubá, Brazil
[2]Federal University of Minas Gerais, Brazil

Introduction

The purpose of this study was to quantitatively assess occupational exposure to noise and whole-body vibration (WBV), analyze the possibility of an association between these two hazards in causing noise-induced hearing loss (NIHL), and discuss feasible means of prevention.

Methods

This cross-sectional study was conducted on a population of 141 bus drivers in São Paulo, Brazil. The prevalence of NIHL was established through audiometric examinations of all workers. The audiograms were classified according to audiometric criteria formulated by Merluzzi et al. [1979]. The entire group of bus drivers was classified and stratified internally in subgroups of "exposed" and "controls" according to the cumulative length of working time as a bus driver. A questionnaire was designed and applied to retrieve occupational history and other relevant information. The methodology for noise exposure evaluation was based on the procedures established in ISO 1999 (1990), as well as in the European Economic Community Directive [EEC 1986]. The methodology for whole-body vibration (WBV) exposure evaluation was based on the principles established in ISO 2631 (1985). The vibration meter used did not have the ISO 2631 (1997) configuration available at the time. Furthermore, the ISO 2631 (1997) application still has many controversial points. As Griffin [1998] pointed out: "International Standard 2631 (1997) is unclear in several important areas." The association between the dependent variable and the set of explanatory variables was analyzed through nonconditional multivariable logistic regression.

Results

It was found that drivers of front-engine buses were exposed to noise levels of 83.6 ± 1.9 dB(A) (weekly average), while drivers of rear-engine buses were exposed to noise levels of 77.0 ± 1.1 dB(A). Also, the average of noise exposure equivalents (Leq) for drivers of front-engine buses was 83.1 ± 1.9 dB(A), whereas the average Leqs for drivers of rear-engine buses was 76.2 ± 1.7 dB(A). This difference was statistically significant. The average vibration acceleration found in the five different bus types analyzed, weighted according to the number and length of use of vehicles, was 0.85 m/s^2. Two factors could explain this high value: the pavement quality of the streets and the suspension type (parabolic suspension springs) used by the bus analyzed. The NIHL prevalence rates were 46% and 24% for the exposed group and the reference group, respectively. It was found that the odds ratio for NIHL prevalence, according to the univariable analysis, was 2.68 (1.31–5.48, for a 95% confidence interval) for the exposed group compared to the reference group. For the best adjusted model, the multivariable analysis showed that age (44 years and older), reference to diabetes, and level of noise exposure (above 86.8 dB(A)) were risk factors relevant to the development of NIHL. Exposure to WBV, when represented by doses, was not significant for NIHL development.

Conclusions

The bus drivers analyzed in this cross-sectional study were exposed to relevant WBV levels, higher than the permissible exposure levels established by ISO 2631 (1985). Also, the noise exposure levels found at the drivers' work station in front-engine buses were higher than those observed at the drivers' work station in rear-engine buses. No association between WBV exposure and NIHL prevalence was observed in this study. In addition, this study could not find any interaction between WBV exposure and noise exposure. However, further studies are required because other logistic regression models were analyzed showing that an interaction between WBV and NIHL could occur, thus increasing the potential risk of adverse health effects to workers.

References

EEC (European Economic Community) [1986]. Council Directive 86/188/EEC of 12 May 1986 on the protection of workers from the risks related to exposure to noise at work. Official Journal L 137, May 24, 1986, pp. 28–34.

Griffin MJ [1998]. A comparison of standardized methods for predicting the hazards of whole-body vibration and repeated shocks. J Sound Vibration *215*(4):883–914.

ISO [1990]. Acoustics – Determination of occupational noise exposure and estimation of noise-induced hearing impairment. Geneva, Switzerland: International Organization for Standardization. ISO 1999:1990.

ISO [1997]. Mechanical vibration and shock: evaluation of human exposure to whole-body vibration. Part 1: General requirements. Geneva, Switzerland: International Organization for Standardization. ISO 2631-1:1997.

Merluzzi F, Parigi G, Cornacchia L, Terrana T [1979]. Metodologia di esecuzione del controllo dell'udito dei lavoratori esposti a rumore (in Italian). Nuovo Arch Ital Otol *7*(4):695–714.

THE COMPENSATORY CONTROL MODEL AS A FRAMEWORK FOR UNDERSTANDING WHOLE-BODY VIBRATION

R. Hancock,[1] N. J. Mansfield,[1] V. K. Goel,[2] and J. Vellani[1]

[1]Environmental Ergonomics Research Centre, Loughborough University, Loughborough, U.K.
[2]Indian Institute of Technology Roorkee, India

Introduction

There is currently a lack of detailed models of the cognitive effects of whole-body vibration (WBV) [Conway et al. 2007], although some general mechanisms are understood. Previous research has often focused on how performance decrements occur. However, a more valuable goal would be to specify why these effects occur and find the importance of the component factors that cause these decrements.

This paper considers Hockey's [1997] compensatory control model (CCM) as a framework for explaining the effects of WBV on activities. The CCM accounts for the hidden costs and stressors resulting from maintaining performance as conditions become more difficult. The CCM allows for a possible explanation of why objective measures of performance can remain stable while workload increases (Figure 1). Below an external load threshold, coping occurs unconsciously (loop A), but active coping is required for elevated external loads, such as WBV.

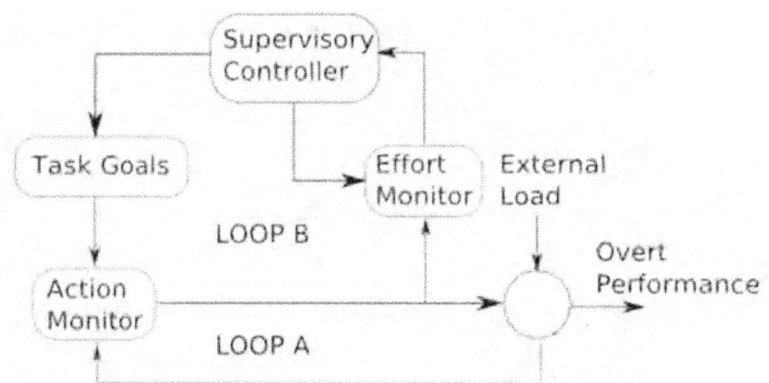

Figure 1.—Compensatory control model (adapted from Hockey [1997]).

The CCM states that performance protection may occur as difficulty and environmental disturbances increase, but at a cost to the performer (e.g., increased psychological stress and fatigue); this is known as active coping. This increased effort is maintained by heightened motivation, but if these costs become too great, the motivational strategy will change to minimize the effects of long-term stress. The simple methods used in previous experiments to determine performance detriments may not be sufficient to capture these underlying stress effects. A framework that predicts these stressors is needed. This framework also needs to take into account that performance can compete with other motivational goals (e.g., vigilance; Dorrian et al. [2007]).

Case Studies

Three case studies are used to illustrate this active coping approach. All are experimental studies by the research team, using WBV as a stressor and measuring workload using NASA-TLX (task load index), a standard technique using six multidimensional scales including mental demand, physical demand, temporal demand, performance, effort, and frustration [Hart and Staveland 1988]. In case study 1, Hancock et al. [2008] reported the effects of WBV magnitude (triaxial random, 0.35, 0.6, 0.87 ms^{-2} rss) and posture (working on table/lap) on subjective difficulty of reading and writing in Hindi and English. Subjective workload scores increased with vibration magnitude. The greatest workload increases were on the frustration and effort subscales. While physical and temporal demand also increased, there was little increase in perceived performance rating. Thus, subjects were aware of increased workload associated with their active coping due to WBV load.

In case study 2, Newell and Mansfield [in press] considered the influence of WBV and twisted postures on reaction time and error rate for a choice reaction time task. "With armrest" conditions showed little or no differences in objective measures of performance with and without WBV (x: 1.4; z: 1.1 ms^{-2}). However subjective workload scores increased, particularly in the effort and frustration subscales. Thus, performance was maintained at the cost of higher mental workload. Without armrests, performance degraded with WBV; this was associated with increased workload. Here, the active coping was insufficient to maintain objective performance.

In case study 3, the influence of WBV (x: 1.4; z: 1.1 ms^{-2}) and 10° seat inclination on joystick performance was considered. It was found that subjective workload increased between WBV exposure and no exposure, although reaction time did not increase. This again suggests that for performance to be maintained, psychological costs are incurred, supporting the CCM.

Summary

Three case studies show that objective measures of performance can remain constant while subjective workload increases. This supports the CCM, where performance is protected at cost to other systems. In particular, fatigue is shown to be one cost induced by active coping. This may reduce the vigilance required to maintain task performance [Dorrian et al. 2007].

Acknowledgment: This research was funded by the EU Asia-Link ASIE/2005/111000 CIRCIS (Collaboration in Research and Development of New Curriculum in Sound and Vibration).

References

Conway GE, Szalma JL, Hancock PA [2007]. A qualitative meta-analytic examination of whole-body vibration effects on human performance. Ergonomics *50*(2):228–245.

Dorrian J, Roach GD, Fletcher A, Dawson D [2007]. Simulated train driving: fatigue, self-awareness and cognitive disengagement. Appl Ergon *38*(2):155–166.

Hancock R, Mansfield NJ, Goel VK, Narayanamoorthy R [2008]. Influence of vibration on workload whilst reading and writing on trains. In: Proceedings of the Ergonomics Society.

Hart SG, Staveland LE [1988]. Development of NASA-TLX (task load index): results of empirical and theoretical research. In: Hancock PA, Meshkati N, eds. Human mental workload. Amsterdam, Netherlands: North Holland B.V., pp. 139–183.

Hockey GRJ [1997]. Compensatory control in the regulation of human performance under stress and high workload: a cognitive-energetical framework. Biol Psychol *45*(1–3):73–93.

Newell GS, Mansfield NJ [in press]. Evaluation of reaction time performance and subjective workload during whole-body vibration exposure while seated in upright and twisted postures with and without armrests. Int J Ind Ergon.

Session VI: Human Body Modeling and Vibration II

Chair: David Wilder
Co-Chair: Arild Øvrum

Presenter	Title	Page
S. K. Patra Concordia University	Defining Reference Values of Vertical Apparent Mass of Seated Body Exposed to Vibration	57
H.-J. Kim University of Michigan	Transmissibility of Upper-Body Segments Affecting Reaching Tasks Under Vehicle Vibration Exposure	59
A. Pranesh Concordia University	Analysis of Distributed Responses of the Seated Human Body Exposed to Vertical Seat Excitations	61
R. Blood University of Washington	A Comparison of Whole-Body Vibration Exposures Across Three Different Types of Bus Seats	63
B. Valero University of Illinois at Chicago	Ride Comfort Evaluation Using a Three-Dimensional Model of a Human Body Focusing the Lumbar Spine Area	65

DEFINING REFERENCE VALUES OF VERTICAL APPARENT MASS OF A SEATED BODY EXPOSED TO VIBRATION

S. K. Patra,[1] S. Rakheja,[1] P.-É. Boileau,[2] H. Nelisse,[2] and J. Boutin[2]

[1]Concordia Centre for Advanced Vehicle Engineering (CONCAVE), Concordia University, Montreal, Quebec, Canada
[2]Institut de recherche Robert-Sauvé en santé et en sécurité du travail (IRSST), Montreal, Quebec, Canada

Introduction

The ranges of biodynamic response of the human body seated without a back support and exposed to vertical whole-body vibration have been standardized in ISO 5982 (2001) in terms of driving-point impedance and apparent mass (APMS). The ranges were derived from synthesis of data reported in different studies, where the total body mass of the subjects varied from 49 to 93 kg, with mean body mass of 75 kg. The standard also provides reference values for three body masses (55, 75, and 90 kg) derived from a linear model. The defined mean values are often interpreted to correspond to mean body mass of 75 kg (DIN standard 45676), which is questionable. Furthermore, the lower and upper limits of the idealized range do not represent the biodynamic response of any particular body mass group. Many studies [Wang et al. 2004] have shown that APMS response of a seated human is strongly influenced by the body mass, excitation magnitude, and sitting posture. It is evident that reliable reference data for particular body masses and sitting postures involving back support have not been defined for development of effective anthropodynamic manikins. This study involves with the characterization of the biodynamic responses of seated human subjects for particular body masses, sitting postures, and excitation magnitudes in order to define the reference values.

Methods

Experiments were undertaken to measure the APMS responses of seated human subjects of particular body masses under vertical vibration. Subjects with body mass in the vicinity of 55, 75 and 98 kg were recruited for the experiments. A total of 27 adult male subjects (9 subjects in each group) participated in the study. Attempts were made to recruit test subjects with body mass close to one of the chosen target masses. The APMS characteristics of seated human subjects were investigated using a Whole-Body Vehicular Simulator. Measurements were performed with each subject seated on a rigid seat without a back support (NBS) and with an inclined rigid backrest (WBS) with an inclination of 12° with respect to the vertical axis. The postural variations caused by two different hand positions (hands in lap (LP) and hands on steering wheel (SW)) were considered only for the WBS condition. The simulator was programmed to synthesize random vertical vibration with flat acceleration power spectrum in the 0.4–20 Hz frequency range with three different magnitudes: 0.5, 1, and 2 m/s^2 root-mean-square (rms) acceleration. The total static and dynamic force of the seated human subject together with the seat was measured using the force platform, while a single-axis accelerometer was installed in the middle of the base platform of the seat to measure acceleration due to vertical excitation. The measured data were appropriately corrected for the rigid seat inertia force, and the APMS characteristics of the seated human subject were derived using the 50-Hz bandwidth and the frequency resolution of 0.0625 Hz. The coherence between the two signals was also monitored during the experiments, which was observed to be close to unity for all of the measurements.

Results and Discussion

The APMS responses of each subject attained during the three trials were analyzed to derive the mean magnitude and phase responses for each mass group, back support condition, and excitation magnitude. The data generally revealed excellent repeatability of measurements, with peak intrasubject variability in the order of 10% in the vicinity of the resonant frequency, which was attributed to relatively small variations in the body mass. Figure 1 illustrates the mean APMS magnitude responses of individual subjects within the 55-kg mass group corresponding to NBS-LP posture under 0.5-m/s^2 excitation. The result showed considerable scatter in the magnitude responses irrespective of excitation level, back support condition, and hand position. The scatter in the magnitude response at lower frequencies can be attributed to slight variations in the body mass within each mass group. Such low-frequency variations may be suppressed by normalizing the biodynamic response data (Figure 2). The normalization, however, cannot eliminate the variabilities around the resonant frequencies, which are attributed to strong body mass effects.

The results, in general, showed strong effects of back support and excitation magnitude, apart from the body mass. Moreover, the effects seem to be strongly coupled. The peak APMS magnitude varied only slightly with the change in the excitation magnitude irrespective of the body mass group. The variation in the primary frequency was observed to be dependent on the back support condition. While the primary resonance frequency decreased with increasing excitation magnitude for the NBS-LP posture, irrespective of the body mass group, the effect diminished for the back-supported posture with hands on steering wheel (WBS-SW). From the results, it was concluded that the reference values of the vertical biodynamic responses need to be defined for particular body mass, sitting posture, and excitation condition. This study further proposes such reference values.

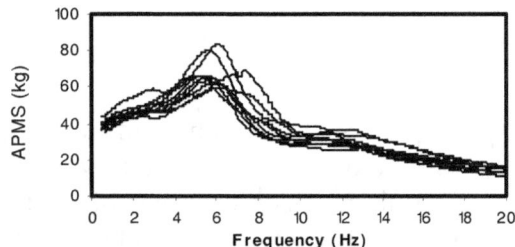

Figure 1.—Measured APMS responses of seated human subjects (NBS-LP) excited under 0.5-m/s^2 rms acceleration.

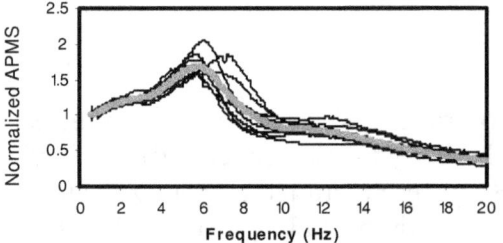

Figure 2.—Normalized APMS responses of seated human subjects (NBS-LP) excited under 0.5-m/s^2 rms acceleration.

References

DIN [1992]. Mechanical impedances at the driving point and transfer functions of the human body. Deutsches Institut für Normung e.V. (German Institute for Standardization). DIN 45676.

ISO [2001]. Mechanical vibration and shock: range of idealized values to characterize seated-body biodynamic response under vertical vibration. Geneva, Switzerland: International Organization for Standardization. ISO 5982:2001.

Wang W, Rakheja S, Boileau P-É [2004]. Effects of sitting postures on biodynamic response of seated occupants under vertical vibration. Int J Ind Ergon *34*(4):289–306.

TRANSMISSIBILITY OF UPPER-BODY SEGMENTS AFFECTING REACHING TASKS UNDER VEHICLE VIBRATION EXPOSURE

Heon-Jeong Kim and Bernard J. Martin
Human Motion Simulation Laboratory, Center for Ergonomics, University of Michigan,
Ann Arbor, MI

Introduction

Vibration perturbation generated by vehicles or terrains is transmitted to the whole body of the seated driver and eventual other operators, thus causing discomfort and interfering with driving vehicles or performing other tasks in dynamic environments. More specifically, vehicle vibration can affect the speed and accuracy of reaching and manipulating tasks associated with vehicle operations. Several studies have investigated the effects of vibration on the seated human with the goal of improving comfort, safety, or manual performance. The apparent mass of the upper torso in static seated postures was analyzed as a function of vibration conditions [McLeod and Griffin 1989; Griffin 1990; Paddan and Griffin 2002]. Reach kinematics and performance under whole-body vibration (WBV) were quantified by the end-effector trajectory, movement time, and aiming error [Reed et al. 2003]. However, to evaluate and predict the effects of WBV on reach performance, vibration transmissibility along the path of body segments must also be identified since upper-body movement coordination can affect the control of the hand or end-effector. This study investigated changes in the transfer functions of individual upper-body segments as a function of various vibration conditions to evaluate the effects of WBV on dynamic reach performance.

Methods

A reaching experiment was conducted on the U.S. Army's Ride Motion Simulator. Three vibration frequencies (2, 4, and 6 Hz) and two vibration directions (vertical and fore-and-aft) were used as input variables to generate six sinusoidal vibration conditions. Thirteen participants performed self-paced reaches to five targets representing the overall right-reach hemisphere for in-vehicle operations. Twenty-six retroreflective markers were placed on selected body landmarks to record reach movements of seated operators. An optical motion capture system (Vicon) was used to acquire movement data with a sampling frequency of 100 Hz. A frequency analysis was used to determine the transfer function of body segments, focusing specifically on the motion of the torso, right upper arm, and right lower arm-hand. The motions of these three body segments were quantified by measuring the displacements of the right shoulder, elbow, and fingertip. Relative displacements of individual upper-body segments were transformed into transfer functions of the right shoulder, right elbow, and right fingertip in the frequency range of 0.2–15 Hz, corresponding to the bandwidth of interest.

Results and Discussion

The transfer functions of right upper-body segments vary as a function of vibration frequency, direction, and spatial target location. For both vertical and fore-and-aft WBV, transmissibility along the right arm shows the same characteristics. Transmissibility is amplified from the shoulder to the fingertip for the 2-Hz vibration, whereas it is attenuated for the 6-Hz vibration. For the 4-Hz vibration, all transfer functions have similar peak values. However, the peak frequency of transmissibility (transfer function) for each body segment varies as a function of vibration direction (Figure 1). For the vertical vibration, the peak of transmissibility occurs at different vibration frequencies for the different body segments. For the fore-aft vibration, all the peaks of transmissibility decrease monotonously as WBV frequency increases. Transmissibility of body segments may also be affected by target location, which is associated with posture differences and movement coordination. Further investigation of transfer functions will be used to develop an active biodynamic model to predict human behavior in WBV environments and to design controls or interfaces adapted for in-vehicle operations.

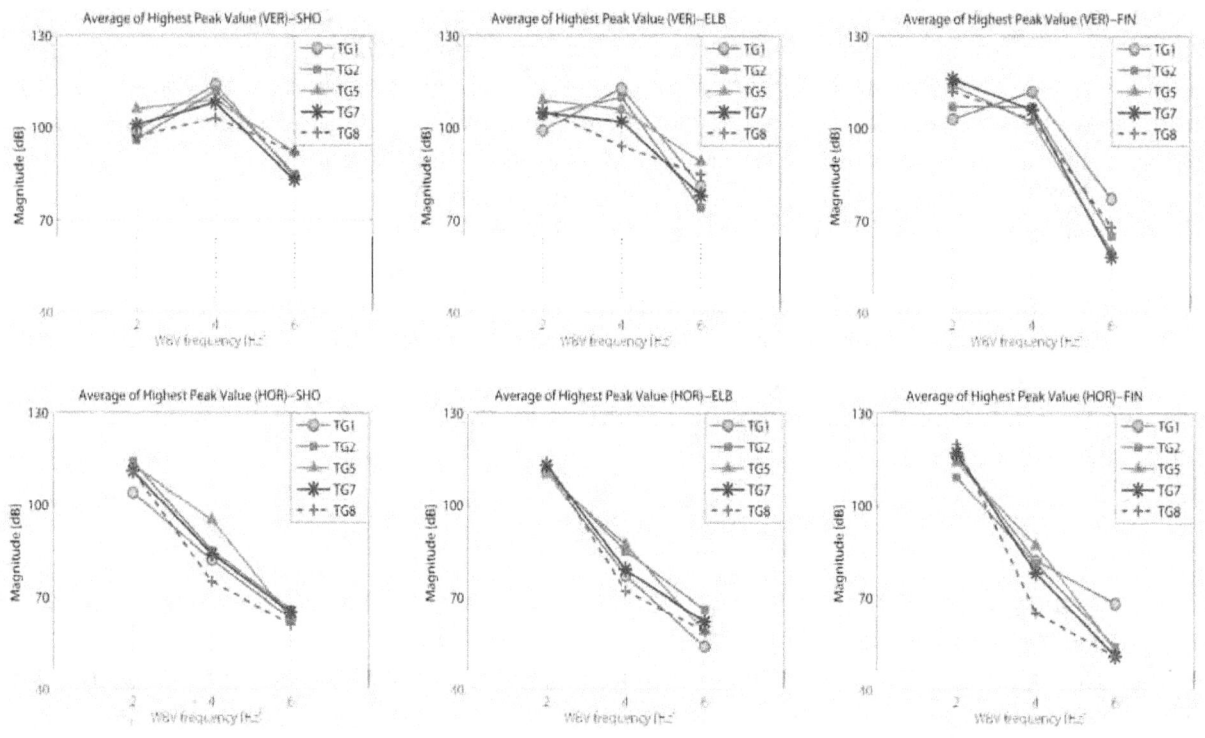

Figure 1.—Average peaks of body segment transfer functions as a function of WBV frequency and direction: shoulder (1st column), elbow (2nd column), and fingertip (3rd column); vertical WBV (1st row) and horizontal WBV (2nd row).

References

Griffin MJ [1990]. Handbook of human vibration. London: Academic Press.

McLeod RW, Griffin MJ [1989]. A review of the effects of translational whole-body vibration on continuous manual control performance. J Sound Vibration 133(1):55–115.

Paddan GS, Griffin MJ [2002]. Effect of seating on exposures to whole-body vibration in vehicles. J Sound Vibration 253(1):215–241.

Reed MP, Chaffin DB, Nebel KJ, Rider KA, Mikol KJ [2003]. A pilot study of the effects of vertical ride motion on reach kinematics. Warrendale, PA: Society of Automotive Engineers, Inc., technical paper 2003-01-0589.

ANALYSIS OF DISTRIBUTED RESPONSES OF THE SEATED HUMAN BODY EXPOSED TO VERTICAL SEAT EXCITATIONS

A. Pranesh,[1] S. Rakheja,[1] and R. DeMont[2]

[1]Concordia Centre for Advanced Vehicle Engineering (CONCAVE), Concordia University, Montreal, Quebec, Canada

[2]Department of Exercise Science, Concordia University, Montreal, Quebec, Canada

Introduction

The seated human body's responses to whole-body vibration (WBV) are characterized mainly on the basis of biodynamic functions such as mechanical impedance, apparent mass (APMS), or absorbed power measured at the driving point. Responses have also been measured in terms of vibration transmitted from the seat to the head (STHT) and other locations of the body. It has been suggested that the vibration power absorbed (Pabs) by the human body is associated with the potential for tissue damage and may thus be a better indicator of injury risks [Vibration Injury Network 2001]. This response quantity, however, is generally evaluated from the signals measured at the driving point and may not be associated with localized responses or potential injuries in various substructures of the body. Vibration responses measured non-invasively at different segments of the human body could yield a better understanding of the localized segmental motions under WBV. Alternatively, anthropometric biomodels, developed with adequate consideration of the anatomical substructures and reliable inertial and joint parameters, could be applied for predicting the segmental motions and interactions among the various segments. Analytical methods could further facilitate the derivation of response quantities that are at present inaccessible by noninvasive techniques, such as the distributed absorbed power responses. In this study, a multibody model of the seated human exposed to vertical excitation was developed to derive the whole-body distributed biodynamic responses in terms of the power absorption characteristics.

Methods

An anthropometric model of the seated human body was constructed using 14 rigid bodies coupled through joints and force elements to permit vertical translations and sagittal-plane rotations of the individual bodies. The upper-body segments (head, neck, thorax, lumbar, torso, abdominal viscera, and sacropelvic element), along with upper and lower limbs, were coupled by appropriate translational and rotational viscoelastic joints. Inertial properties and joint coordinates were based on a 50th-percentile adult male sitting in a passenger's posture with "hands in lap." Stiffness and damping values for the viscoelastic elements were identified through minimization of the errors between the model's biodynamic responses and experimentally measured target values. The target values were based on the laboratory-measured APMS and STHT responses of six adult male subjects (body mass: 72–79.6 kg) exposed to random vertical vibration (1 m/s^2 rms; 0.5–15 Hz) [Wang 2006]. A gradient-based constrained error minimization technique was adopted in the MSC.ADAMS platform, with the model being simulated under a vertical swept harmonic excitation at the seat. It was concluded that STHT serves as a better target data for anthropometric model parameter identification than the APMS. Also, modal properties of the former compared better with those reported in the literature. The model derived on the basis of STHT error minimization alone was thus used to analyze segmental vibration responses. The total and distributed vibration power absorption responses were computed from the energy dissipation across the dissipative components at selected discrete frequencies. Unlike the global power responses generally derived from the driving point responses, the model computes the distributed power responses from the relative velocities of different body segments occurring at the joints, which obviously are not directly measurable. The estimated power responses across the individual joints are normalized by the mass/inertia supported by the joint.

Results

In Figure 1A, the total Pabs response of the model is compared with the mean measured data reported by Wang [2006] for the same six subjects used for target data set generation. The model shows acceptably good agreement with the measured Pabs for the entire frequency range, except for small deviations in the vicinity of the primary resonance. The results suggest that the model can predict the global absorbed power response reasonably well. Figure 1B shows the distributed energy dissipation characteristics of the viscous joints between model segments in vertical translation, normalized with the total mass supported by the corresponding joint. The results suggest highest energy dissipation within the visceral tissue, followed by the buttock, lower lumbar, and neck joints.

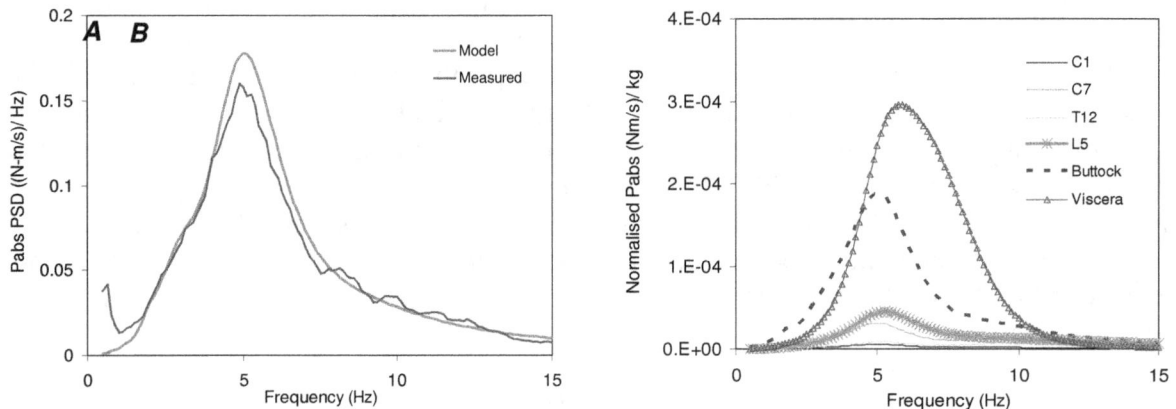

Figure 1.—A, spectral density of total absorbed power; B, model: distributed energy dissipation.

Discussion

The anthropometric model developed in this study is capable of effectively predicting body segment responses to vertical WBV. The derivation of distributed responses, especially in absorbed power, is also definitely possible with this approach. These response variables may yield significant insights into potential injury mechanisms in the seated human body due to vibration exposure. However, further efforts are needed to achieve improved confidence in the model's predictions.

References

Vibration Injury Network [2001]. Review of methods for evaluating human exposure to whole-body vibration. Appendix W4A to final report. [http://www.humanvibration.com/EU/VINET/pdf_files/Appendix_W4A.pdf]. Date accessed: May 2008.

Wang W [2006]. A study of force-motion and vibration transmission properties of seated body under vertical vibration and effects of sitting posture [Dissertation]. Montreal, Quebec, Canada: Concordia University.

A COMPARISON OF WHOLE-BODY VIBRATION EXPOSURES ACROSS THREE DIFFERENT TYPES OF BUS SEATS

Ryan Blood, Jim Ploger, and Peter W. Johnson
University of Washington, Department of Environmental and Occupational Health Sciences, Seattle, WA

Introduction

Low back pain among public transit drivers is quite prevalent. U.S. Bureau of Labor Statistics data indicate a high occurrence of low back illness and injury in the transportation industry. The relationship between whole-body vibration (WBV) exposure and low back pain has been established in prior research [Troup 1998]. Using a low-floor coach bus and a standardized test route, the goal of this project was to compare and determine whether there were differences across three seat types in their ability to attenuate WBV exposures as defined in ISO 2631-1 (1997) and ISO 2631-5 (2004). By using a standardized and controlled test route, which accurately simulates on-the-job conditions, it is hoped that the analysis of these data may help direct and reduce WBV exposures and the potential for subsequent injuries and illnesses among bus operators.

Methods

Sixteen subjects were recruited for this study. After data collection (dropouts and equipment failure), complete repeated measures data were collected and analyzed from 10 subjects. Three different seats were used: (1) a Recaro Ergo M, (2) a USSC Q91 with a standard foam seat pad, and (3) the identical USSC Q91 retrofitted with a silicone foam seat pad. Subjects drove the bus for approximately 1 hr over a 65-km standardized test route. Seat order was not randomized. Vibration was measured using two triaxial accelerometers, one mounted on the seat using an accelerometer rigidly mounted in a runner seat pad and the other placed on the floor next to the seat using thin high-bond adhesive. Based on ISO 2631-5, which requires a sampling rate of at least 160 samples per second, we collected data at 640 samples per second. The standardized test route included surface streets, freeways, and a small section of road containing eight speed humps. This route was chosen to represent different types of driving, including start-and-stop driving associated with surface streets, impulsive speed hump excursions, and continuous freeway travel. The bus used in the study was a 4-year-old, empty, 12.2-m, low-floor coach bus (New Flyer Industries, Inc., Winnipeg, Manitoba, Canada). Since there are no steps to impede entrance and exit, most major metropolitan bus companies are transitioning to this type of bus.

The instrumentation developed for this study included a PDA-based portable WBV data acquisition system, which collected raw unweighted WBV data at 640 Hz, and the associated software to analyze WBV exposures, per ISO 2631-1 and 2631-5. The preliminary analysis of the data was focused on analyzing the Z-axis measurements of A_w, vibration dose value (VDV), crest factor, maximum continuous peak, and positive and negative raw peaks. Data are presented as mean and standard error with significance accepted for p-values less than 0.05.

Results

The global positioning system data indicated that there were no significant differences in bus speed across the three conditions. Table 1 shows the preliminary results of the analysis of time-weighted average (TWA) and peak data across the three different seat types and between the bus seat and floor. Relative to the vibration measured at the floor, the bus seats primarily attenuated the vibration exposures. The degree of attenuation depended on the type of measure; average measures of vibration (A_w and VDV) were attenuated to a lesser degree than impulsive measures (crest factor, maximum peak, and raw peaks). In addition, with the exception of VDV, the vibration measured on the bus floor was dependent on seat type. There were also differences across seats in the attenuation of the impulsive vibration measures (crest factor, maximum peak, and raw positive peak). In general, the Recaro seat had lower impulsive exposures, but what the data do not show (not presented) are the seat results by individual road types. On freeways, relative to floor measurements, all seats amplified the A_w exposures; over the speed humps, all seats amplified the VDV exposures.

Table 1.—Mean and standard error Z-axis vibration measures by location and seat type (n = 10)

	Seat				Floor			
	Recaro Ergo M	USSC Q91	USSC Q91 with silicone	p-value	Recaro Ergo M	USSC Q91	USSC Q91 with silicone	p-value
A_w (m/s^2)	0.39 ± 0.01	0.39 ± 0.02	0.39 ± 0.01	0.98	0.43 ± 0.01	0.42 ± 0.01	0.44 ± 0.01	0.04
VDV (m/s$^{1.75}$)	9.03 ± 0.23	9.28 ± 0.46	9.32 ± 0.43	0.77	11.83 ± 0.46	12.05 ± 0.44	11.11 ± 0.20	0.20
Crest factor	9.65 ± 0.51	12.09 ± 0.36	12.63 ± 0.66	0.001	20.92 ± 1.82	22.85 ± 1.61	15.26 ± 0.82	0.01
TWA peak (m/s^2)	3.61 ± 0.18	4.60 ± 0.27	4.78 ± 0.32	0.005	8.99 ± 0.82	9.45 ± 0.70	6.69 ± 0.39	0.04
Raw (+) peak (m/s^2)	5.44 ± 0.27	8.26 ± 1.03	7.33 ± 0.53	0.03	44.58 ± 5.41	53.72 ± 2.32	32.05 ± 3.94	0.01
Raw (−) peak (m/s^2)	−6.50 ± 0.33	−6.91 ± 0.42	−7.08 ± 0.74	0.77	−47.88 ± 6.00	−62.87 ± 4.87	−34.02 ± 4.20	0.01

Discussion

The bus seats tested in this study seem to be attenuating vibration exposures, but there are some exceptions, such as freeways (A_w) and speed humps (VDV). This indicates that the bus seats are not optimized for all road types and perhaps bus seat selection could be improved by matching or tuning the bus seat to the predominant road type on the bus route. Of particular interest is the amplification of exposures on freeways, which can predominate certain types of bus routes. Finally, the vibration measured on the bus floor was dependent on seat type. This indicates a complex interaction between the bus seat and vibration measured on the floor.

References

ISO [1997]. Mechanical vibration and shock: evaluation of human exposure to whole-body vibration. Part 1: General requirements. Geneva, Switzerland: International Organization for Standardization. ISO 2631-1:1997.

ISO [2004]. Mechanical vibration and shock: evaluation of human exposure to whole-body vibration. Part 5: Method for evaluation of vibration containing multiple shocks. Geneva, Switzerland: International Organization for Standardization. ISO 2631-5:2004.

Troup JDG [1988]. Clinical effects of shock and vibration on the spine. Clin Biomech 3(4):232–235.

Zuo L, Nayfeh SA [2003]. Low order continuous-time filters for approximation of the ISO 2631-1 human vibration sensitivity weightings. J Sound Vibration 265(2):459–465.

RIDE COMFORT EVALUATION USING A THREE-DIMENSIONAL MODEL OF A HUMAN BODY FOCUSING ON THE LUMBAR SPINE AREA

Bertrand Valero, Farid Amirouche, and Vincent Aledo
Vehicle Technology Laboratory, University of Illinois at Chicago (UIC)

Introduction

It is estimated that about 8 million to 10 million U.S. workers are exposed daily to whole-body vibration (WBV) [Wasserman 1987]. These workers are exposed to WBV through various sources, e.g., the operation of trucks, forklifts, buses, heavy equipment, farm vehicles, and helicopters. WBV can limit the ability of the operator to perform his or her tasks, which can lead to safety issues. Vibration can enter the human body through different pathways, e.g., via the spine while driving a vehicle. In the case of heavy vehicle operators, long-term exposure to WBV has been linked to severe low back pain in the lumbar region [Schwarze et al. 1998]. A higher rate of lumbar syndrome exists among people who have been subjected to this type of vibration for an important period of their life.

Suspension systems have been introduced between the seat and the chassis of such vehicles to minimize body vibration. In parallel to the research done at the UIC Vehicle Technology Laboratory on the design of an active seat suspension [Valero et al. 2007], a three-dimensional model of the human body is being developed to evaluate the absorbed power and energy transmitted to the human body. The model aims to assert the performance of seat design using the latter measures.

Methods

A human body model including 19 segments and 18 joints was created. The inertia properties of each segment were defined using anthropometric data for a 95th-percentile American male. The properties of a Hybrid III crash test dummy were initially used for the joint characteristics between the segments to build the model. Then, using experimental results found in the literature, the joint characteristics were tuned to obtain a first human body resonance comparable to the one observed experimentally when the seat-to-head transmissibility was considered.
A finite-element model of the lumbar spine developed in our laboratory and validated against experimental data [Varadarajan et al. 2002] was used to replace the segment representing the lower torso.

The seat suspension design developed in the Vehicle Technology Laboratory was fixed on a shaker table. The acceleration signals of the seat and table were collected during the testing. These experimental data were applied to a body representing the seat and connected to the human body model through the pelvis. The interaction between the body and the seat was modeled through a spring and damper in parallel [Amirouche et al. 1997]. The absorbed power and the energy transmitted to the body could then be estimated.

Results

Experimental results of the semiactive suspension system show a significant reduction in the root-mean-square value of the acceleration of the seat. A reduction of the total absorbed power by the operator is expected and will allow us to better assess the performance of the suspension design.

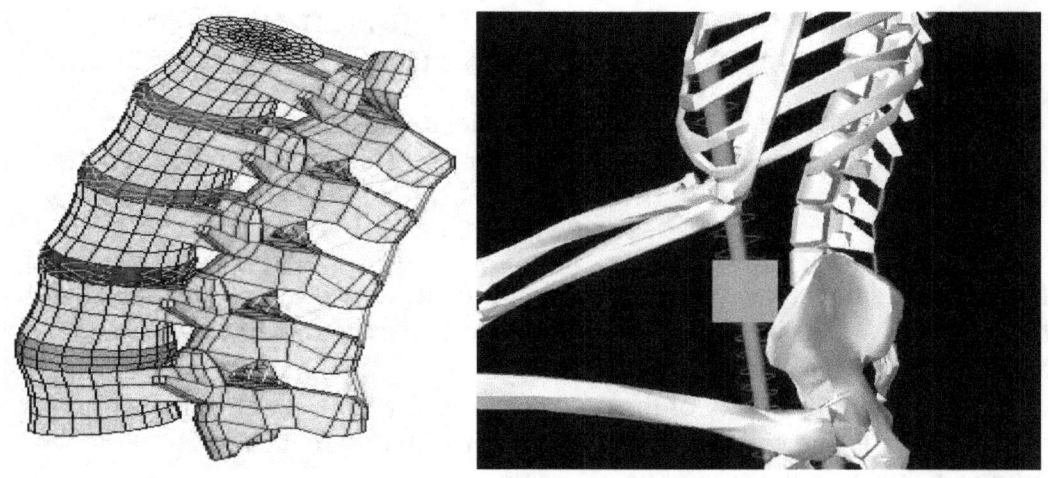

Figure 1.—Human body model.

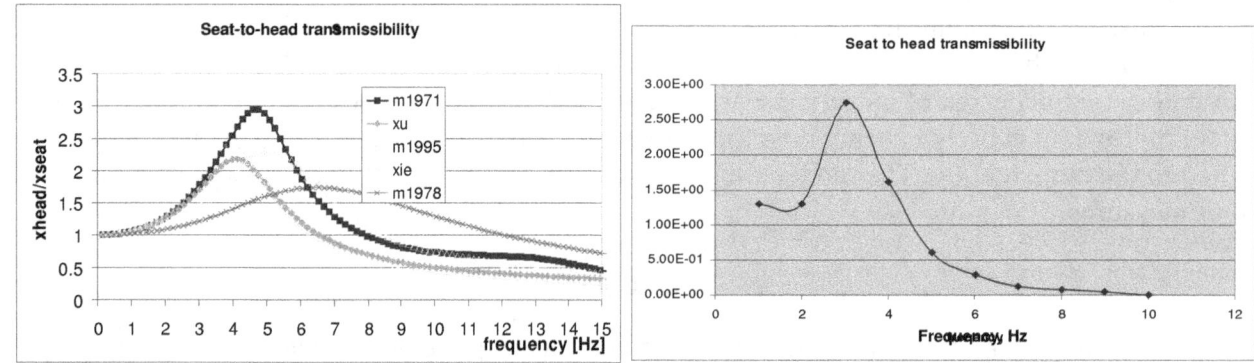

Figure 2.—Seat-to-head transmissibility obtained through different models *(left)* and seat-to-head transmissibility given by our model including the lumbar spine *(right)*.

References

Amirouche F, Alexa E, Xu P [1997]. Evaluation of dynamic seat comfort and driver's fatigue. Warrendale, PA: Society of Automotive Engineers, Inc., technical paper 971573.

Schwarze S, Notbohm G, Dupuis H, Hartung E [1998]. Dose-response relationships between whole-body vibration and lumbar disk disease: a field study on 388 drivers of different vehicles. J Sound Vibration *215*(4):613–628.

Wasserman D [1987]. Human aspects of occupational vibration. Amsterdam, Netherlands: Elsevier Publishers.

Valero B, Amirouche FML, Mayton AG, Jobes CC [2007]. Comparison of passive seat suspension with different configuration of seat pads and active seat suspension. Warrendale, PA: Society of Automotive Engineers, Inc., technical paper 2007-01-0350.

Varadarajan R, Amirouche FML, Wagner F, Guppy K [2002]. Effect of osteoporosis in a disc degenerated lumbar spine. In: Proceedings of the Fourth World Congress of Biomechanics (Calgary, Alberta, Canada, August 4–9, 2002).

Session VII: Hand-Arm Vibration

Chair: Ren Dong
Co-Chair: Richard Stayner

Presenter	Title	Page
U. Kaulbars BGIA	Hand-Arm Vibration Exposure in Aircraft Manufacture: Measures of Vibration Attenuation	68
T. McDowell National Institute for Occupational Safety and Health (NIOSH)	Vibration Transmitted From an Impact Wrench to the Human Wrist and Elbow	70
J. Kim University of Cincinnati	Development of a Receptance Method-Based Modeling Technique for Finite-Element Analysis of Hand-Arm Vibration	72
A. Turcot * Institut national de santé publique du Québec (National Institute of Public Health of Québec)	Noise and Hand-Arm Vibration: A Dangerous Match	74
S. A. Adewusi Concordia University	Posture Effect on Vibration Transmissibility of the Hand-Arm	76
R. Dong NIOSH Morgantown	A New Biodynamic Approach to Assess the Effectiveness of Antivibration Gloves	78

* Author unable to attend conference.

HAND-ARM VIBRATION EXPOSURE IN AIRCRAFT MANUFACTURE: MEASURES OF VIBRATION ATTENUATION

Uwe Kaulbars
BG-Institute for Occupational Safety and Health (BGIA)
Sankt Augustin, Germany

Introduction

Despite a high level of automation, a large number of handheld power tools are still being used during the manufacture of aircraft. In particular, high exposure to vibration occurs during use of a riveting hammer and dolly. Other tools typically used that give rise to vibration exposure are power screwdrivers, saws, and grinding machines. Under the EU Vibration Directive, employers are obliged to take state-of-the-art measures to reduce vibration exposure when the daily vibration exposure A(8) exceeds 2.5 m/s^2.

From the range of measures within a vibration attenuation program (as described in the CR 1030 series of CEN reports), this work studied the use of alternative working methods and the selection of suitable working equipment that could lower vibration exposure.

Methods

Measurements were performed under typical work and plant conditions. Each measurement was repeated several times with a number of different test persons experienced in the use of the equipment. For riveters, a working cycle of 10 riveting operations was specified for each discrete measurement in order to permit comparison between the different tools for a given task. Measurement and analysis of the results were conducted in accordance with ISO 5349.

Results

The frequency-weighted acceleration (a_{hw}) values are between 5.4 and 11.5 m/s^2 for conventional riveting hammers without vibration damping and between 1.75 and 6.8 m/s^2 for riveting hammers with vibration damping. These values are not conclusive, however, since they are determined from the number of riveting operations over the integration period (duration of measurement) according to the hardness of the rivet and the individual pace of progress. The number of riveting operations that can be performed before the daily vibration exposure A(8) of 2.5 m/s^2 is exceeded is 530–1,170 per day for a riveting hammer without damping and 1,830–3,000 for a riveting hammer with damping. This takes into account any breaks necessitated by the work.

Higher exposures occur on the dolly. Here the daily vibration exposure is exceeded after 98 riveting operations per day. The vibration exposure can be reduced by increasing the mass of the dolly (Table 1). Owing to the confined conditions, the use of dollies with vibration damping is frequently not possible (Figure 1).

Vibration attenuation can be improved considerably by the use of an alternative working method. Blind rivet guns employ an impact-free process; a titanium pop rivet is deformed by means of a drawing movement. Blind rivet guns can perform 25,160 riveting operations per day before the daily vibration exposure A(8) of 2.5 m/s^2 is exceeded. Threaded joints also generate lower vibration exposure than riveted joints.

Table 1.—Reduction in vibration exposure as a function of dolly weight

Dolly	No. of riveting operations per day before $A(8) = 2.5$ m/s² is reached
Tungsten, 3.16 kg	538
Steel, 1.61 kg	293
Steel, 0.839 kg	109

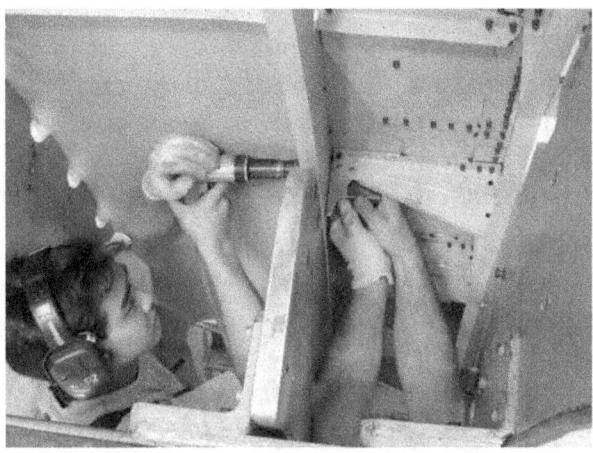

Figure 1.—Use of a riveting hammer and dolly in confined conditions.

Further relevant vibration exposure occurs during the sawing of 2 mm sheet aluminium with a pad saw ($a_{hw} = 6.9 \pm 1.7$ m/s²) and during the use of random orbital sanders ($a_{hw} = 5.1 \pm 1.1$ m/s²).

Summary and Discussion

The effects of vibration attenuation for riveting equipment were investigated. Damping increases the number of riveting operations for riveting hammers roughly by a factor of three. Even more effective for prevention purposes is the use of blind rivet guns, which allow up to 25,160 riveting operations per day before the daily exposure limit is exceeded. This opens alternative options to reduce vibration exposure.

References

EEC (European Economic Community) [2002]. Directive 2002/44/EC of the European Parliament and of the Council of 25 June 2002 on the minimum health and safety requirements regarding the exposure of workers to the risks arising from physical agents (vibration) (16th individual directive within the meaning of article 16(1) of directive 89/391/EEC). Official Journal L 177/13 6.7.2002.

European Committee for Standardization [1995]. Hand-arm vibration: guidelines for vibration hazards reduction – Part 1: Engineering methods by design of machinery. CEN/CR 1030-1:1995.

ISO [2001]. Mechanical vibration: measurement and evaluation of human exposure to hand-transmitted vibration. Part 1: General requirements. Geneva, Switzerland: International Organization for Standardization. ISO 5349-1:2001.

VIBRATION TRANSMITTED FROM AN IMPACT WRENCH TO THE HUMAN WRIST AND ELBOW

X. Xu, D. E. Welcome, T. W. McDowell, C. Warren, and R. G. Dong

Health Effects Laboratory Division, National Institute for Occupational Safety and Health, Morgantown, WV

Introduction

Extensive use of impact wrenches could expose operators to prolonged, intensive vibration. Such exposure could result in hand-arm vibration syndrome. The effects of vibration exposure in a structure of the hand-arm system are likely to be more closely related to the actual vibration power absorption at that specific structure [Dong et al. 2008]. The objective of this study was to characterize the vibration transmitted to the wrist of the operators of impact wrenches.

Methods

Six experienced male operators of impact wrenches participated in the experiment. Each of them used 15 impact wrenches on a simulated work station (Figure 1). For each trial, the subject seated 10 nuts onto plate-mounted bolts in a 30-sec period. Five trials were performed for each tool. Triaxial accelerations at three locations (tool handle, wrist, and elbow) were measured. In an effort to better simulate actual work situations in the field, postures were not controlled, and the subject could use a posture judged to be most comfortable. The six subjects generally adopted one of three postures wherein the tool handle was oriented vertically, horizontally, or at a 45° angle. Figure 1 shows the 45° working posture.

In addition, a vibration transmissibility test was performed on a one-dimensional (1-D) vibration system using a broadband random vibration as the excitation (Figure 2). Operators were instructed to keep the same posture and apply the same grip and push forces to the shaker's instrumented handle as they perceived in the tool test. Whereas the instrumented handle provided the measurement of the input vibration, the accelerometers fixed on the subject's wrist and elbow were used to measure the transmitted vibrations. For each subject, six trials were performed lasting 30 sec each.

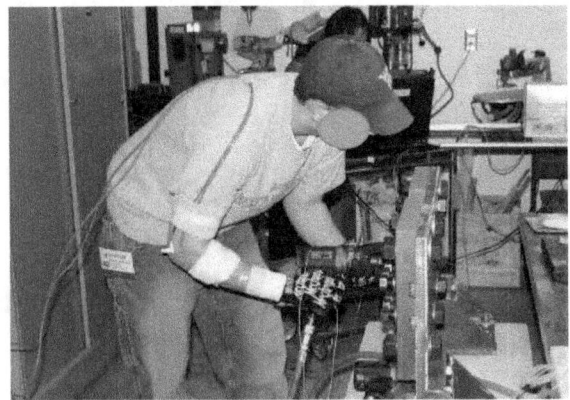

Figure 1.—Operation of an impact wrench.

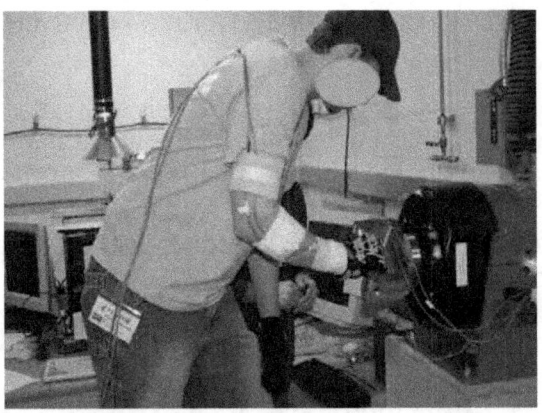

Figure 2.—Measurement of the transmissibility on a 1-D test system.

Results and Discussion

Even though the vibration magnitudes measured at the three locations are statistically significantly different across 6 subjects × 15 tools × 5 trials = 450 trials (ISO-weighted tool handle acceleration: 6.26 ± 2.08 m/s^2; wrist: 7.49 ± 2.54 m/s^2; elbow: 4.05 ± 1.32 m/s^2), they are reliably correlated with each other ($p < 0.001$). The correlation between ISO-weighted tool acceleration and wrist acceleration is shown in Figure 3. Figure 4 shows the transmissibility measured with the six subjects on the 1-D test system, together with ISO frequency weighting (ISOwt), the transmissibility-derived weighting ($W_{Tr\text{-}Wrist}$), and the weighting derived from palm vibration power absorption (VPAwt) [Dong et al. 2008]. The vibration transmissibility shows a resonant peak at 31.5 Hz. A similar trend is observed in the VPA weighting. However, this resonance is not reflected in the ISO weighting.

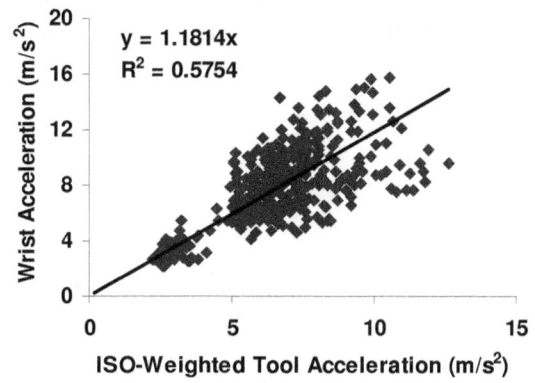

Figure 3.—A correlation relationship.

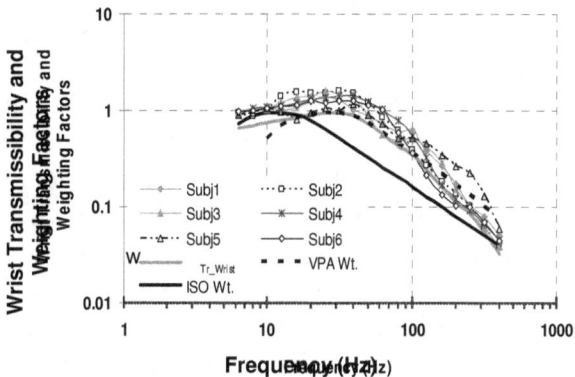

Figure 4.—Vibration transmissibility and frequency weighting.

The reliable correlation between ISO frequency-weighted tool acceleration and the wrist and elbow accelerations suggests that the ISO weighting partially reflects the characteristics of the vibrations transmitted to the wrist and elbow. The results, however, also suggest that the ISO weighting may underestimate the effect of the resonance of the hand-arm system. It is feasible to develop a wrist vibrometer for continuously measuring the vibration transmitted to the wrist. The exposure duration of the vibration can also be accurately quantified with a wrist vibrometer. A method based on wrist vibration measurement may be developed to monitor the exposure and to assess the risk of the exposure at least in the wrist-arm subsystem.

References

Dong JH, Dong RG, Rakheja S, Welcome DE, McDowell TW, Wu JZ [2008]. A method for analyzing absorbed power distribution in the hand and arm substructures when operating vibrating tools. J Sound Vibration *311*:1286–1304.

ISO [2001]. Mechanical vibration: measurement and evaluation of human exposure to hand-transmitted vibration. Part 1: General requirements. Geneva, Switzerland: International Organization for Standardization. ISO 5349-1:2001.

DEVELOPMENT OF A RECEPTANCE METHOD-BASED MODELING TECHNIQUE FOR FINITE-ELEMENT ANALYSIS OF HAND-ARM VIBRATION

J. Kim and S. P. Pattnaik
Department of Mechanical Engineering, University of Cincinnati, Cincinnati, OH

Introduction

Exposure to excessive work-related vibration causes musculoskeletal disorders (MSDs), including abnormal vascular and neurological function. Hand-arm vibration syndrome (HAVS) is a major MSD that often occurs in construction workers and miners. HAVS guidelines have been developed that rely mainly on population study and empirical tests, which provide only limited information. Especially for the hand and fingers, a direct animal test is not a practical option. For this reason, a numerical analysis is very useful in the study of HAVS. This abstract reports on the development of an analysis procedure and related techniques for the study of hand and arm vibration.

A substructuring approach such as the receptance method [Soedel 2004] enables modeling of the hand and arm as a part of the whole body, as shown in Figure 1. Representing appropriate boundary conditions is necessary to implement a receptance method-based analysis. In this abstract, we consider modeling of a finger with suitable boundary conditions to model its dynamic behavior more realistically. Contact analysis of a fingertip using the finite-element analysis (FEA) technique is reported as a preliminary step of receptance method-based hand-arm analysis.

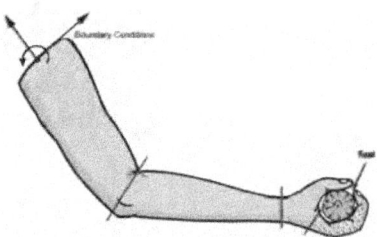

Figure 1.—Hard-arm as a body part.

Methods

Receptance method: The receptance method allows the dynamics of a subsystem modeled independently, then combined with the other system. As the receptance is defined as the ratio of dynamic response and the applied force at the boundary, it is necessary to determine boundary conditions for a part of the hand-arm system model. We begin with developing a finger model with general boundary conditions.

Finger model: We consider a detailed model of a finger, including segments of bone, nail, and subcutaneous tissue. The fingertip considered in Figure 2 is in contact with a linearly elastic PVC polymer probe, which is frictionless. Bone, nail, and probe are modeled by linearly elastic materials by taking the data from Yamada [1970], and soft tissue is modeled by non-linearly elastic and viscoelastic materials by taking the data from Wu et al. [2006a,b]. The Young's moduli of the bone, nail, and probe are 17.0 GPa, 170.0 MPa, and 2.8 GPa, respectively, and Poisson's ratio is 0.30 for all of them.

A commercial FEA software package (Abaqus version 6.7) was used, and a plane strain analysis was conducted. As the fingertip is being pushed with a probe, the other end is held with two sets of springs in either direction. The springs can be modeled to consider the effect of the rest of the hand and body. Figure 2 shows the finite-element model with various segments in different colors and displacement calculated from a contact analysis in response to 2 mm of pad displacement.

Results

The results shown in Figure 2 are the deformation due to 2.0 mm of displacement by probe for a spring-based boundary condition model.

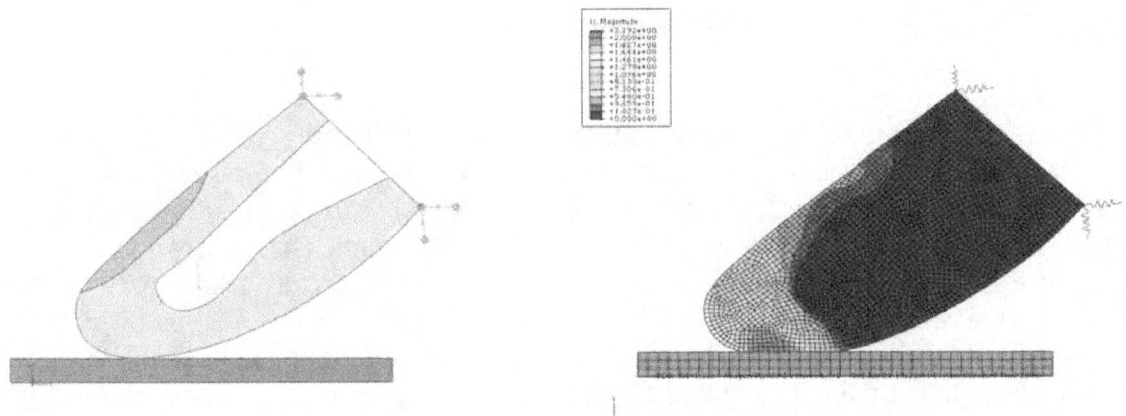

Figure 2.—*Left:* finger model. *Right:* contour plot of the displacement.

Discussion

To validate the receptance method-based model, two sets of assemblies can be compared. For example, in the finger model we may compare:
- Finger + rest of body represented by the receptance method;
- Finger + simple hand + rest of body represented by the receptance method.

Future Directions

The current study will be extended to develop a fundamental tool to study HAVS. Techniques will be developed for finite-element vibration analysis of hand parts, considering the effects of gripping force and constraints coming from the rest of the body. Important response parameters will be identified, and the strategy to calculate them will be developed.

References

Soedel W [2004]. Vibrations of shells and plates. 3rd ed. New York: Marcel Dekker, Inc.

Wu JZ, Dong RG, Welcome DE [2006a]. Analysis of the point mechanical impedance of fingerpad in vibration. Med Eng Phys *28*(8):816–826.

Wu JZ, Welcome DE, Dong RG [2006b]. Three-dimensional finite element simulations of the mechanical response of the fingertip to static and dynamic compressions. Comput Methods Biomech Biomed Eng *9*(1):55–63.

Yamada H [1970]. Strength of biological materials. Baltimore, MD: Williams & Wilkins Co.

NOISE AND HAND-ARM VIBRATION: A DANGEROUS MATCH

Alice Turcot,[1] Richard Larocque,[1] Serge-André Girard,[1] Valérie Roy,[1] and Samuel Tétrault[2]

[1]Institut national de santé publique du Québec (National Institute of Public Health of Quebec), Quebec, Canada
[2]Faculté de médecine (School of Medicine), Laval University, Quebec, Canada

Introduction

Workers exposed to hand-arm vibration generated by vibrating tools are also exposed to high noise levels. Databases supply objective elements regarding not only acceleration levels, but also noise levels. A chain saw can produce noise levels up to 111 dBA, and jackleg drills can reach 115–130 dBA [Umeå University 2008; Boileau et al. 1990]. Studies have shown that forestry workers suffer from both hearing loss and "white finger" [Pyykkö et al. 1989]. A study in Quebec, Canada, showed that 23% of workers' compensation claimants who had white finger were also eligible for compensation based on degree of hearing loss [Turcot et al. 2007]. The study also showed that workers exposed to vibrating tools demonstrated a more severe hearing loss when they also had white finger. This loss is more pronounced in the high frequencies [Pyykkö et al. 1986; Miyakita et al. 1987]. Furthermore, a study demonstrated that hearing loss was about twice as common in men and women who reported finger blanching, including those who had never been exposed to noise and who had never been exposed to hand-transmitted vibration [Palmer et al. 2002]. Quebec has an audiometric exam database conducted for auditory health screening of workers in mobile laboratories under standardized conditions. Our objective was to verify if the hearing loss in noise-exposed workers who have white finger is more pronounced than that of workers exposed to the same noise levels from a critical examination of all audiometric frequencies with recourse to the data of ISO standard 7029, something that studies conducted in the 1980s did not allow. This standard allows valid comparisons with the audiometric exam results of workers exposed to noise and other auditory stress levels (ISO 7029). The preliminary results of our study are presented here.

Methods

From a database of audiometric exams conducted during 1983–1996, 15,751 were retained for analysis (n = 8,528 in mining, and 7,225 in forestry). The following variables were retained: age; sex; noise exposure at previous exam; noise experience (exposure length); gross result to frequencies 0.5, 1, 2, and 4 kHz; and hearing loss associated with noise (gross loss minus normal age loss, as stated in ISO 7029, 50th percentile). Cases of workers affected by white finger disease (WFD) and who received workers' compensation during 1983–1998 were retained for analysis (n = 96). Parametric regression curbs allow the tracking of a workers' hearing profile based on experience with noise. ANCOVA tests allow the verification of auditory differences in WFD-affected and unaffected mining and forestry workers. A literature review on the subject was done in PubMed (1995–2007, French and English articles with the keywords "noise" and "vibration"). A total of 33 articles (animal studies, epidemiology) were retained for analysis.

Results

Among the mining and forestry workers examined, the data showed that 7% received workers' compensation for occupational hearing loss. We observed that the auditory profiles of mining and forestry workers are comparable to workers in other noisy industry sectors.

Secondly, we observed that workers with white finger (n = 96) showed more pronounced hearing loss than mining and forestry workers at the threshold of the 50th percentile of ISO standards. This hearing loss is significant. Loss was noted at the 0.5-Hz frequency and at other frequencies (1, 2, and 4 kHz). Finally, multivariable analysis showed that noise-exposed workers with more than 15 years' seniority have a 1.44 risk of demonstrating significant hearing loss, and workers with white finger have an increased risk of 1.37.

Discussion

Mining and forestry workers examined during screening tests demonstrated occupational hearing loss and were entitled to workers' compensation. This low proportion of cases will be compared with results from other industry sectors. Hand-arm vibration exposure increases the risk of hearing loss in workers affected by WFD. These results agree with the literature. However, our study shows degradation at the low frequencies compared with other studies that show a more pronounced effect at the 4-kHz frequency. Some authors have speculated that vibration may activate the sympathetic nervous system and induce vasospasm in the organ of Corti reflexively. The mechanisms to explain these low-frequency effects are unknown. These results indicate that preventive measures must be put into place for the short term in these industry sectors and reinforce the need to reduce exposure to these two aggressors. Vibration definitely seems to be a potential aggressor for hearing loss, and medical questionnaires must include research on time exposure to noise and hand-arm vibration, as well as white finger cases in industry sectors where workers are exposed to vibrating tools.

References

Boileau P-É, Boutin J, Milette L [1990]. Exposition au bruit et aux vibrations mains-bras liée à l'opération de foreuses à béquille pneumatique et hydraulique (Exposure to noise and hand-arm vibration associated with the operation of pneumatic and hydraulic jackleg drills) (in French). Montreal, Quebec, Canada: Institut de recherche en santé et en sécurité du travail (IRSST). Études et recherches, report R-046.

Miyakita T, Miura H, Futatsuka M [1987]. Noise-induced hearing loss in relation to vibration-induced white finger in chain-saw workers. Scand J Work Environ Health 13(1):32–36.

Palmer KT, Griffin MJ, Syddall HE, Pannett B, Cooper C, Coggon D [2002]. Raynaud's phenomenon, vibration induced white finger, and difficulties in hearing. Occup Environ Med 59(9):640–642.

Pyykkö I, Starck J, Pekkarinen MS [1986]. Further evidence of a relation between noise-induced permanent threshold shift and vibration-induced digital vasospasms. Am J Otolaryngol 7(6):391–398.

Pyykkö I, Koskimies K, Starck J, Pekkarinen J, Färkkilä M, Inaba R [1989]. Risk factors in the genesis of sensorineural hearing loss in Finnish forestry workers. Br J Ind Med 46(7):439–446.

Turcot A, Roy S, Simpson A [2007]. Lésions professionnelles reliées aux vibrations main-bras au Québec, 1993 à 2002 – Partie II: Analyse descriptive des dossiers d'indemnisation des travailleurs (Occupational injuries related to hand-arm vibration in Quebec, 1993 to 2002 – Part II: Descriptive analysis of worker compensation files) (in French). Montreal, Quebec, Canada: Institut de recherche en santé et en sécurité du travail (IRSST). Études et recherches, report R-492.

Umeå University [2008]. Occupational and Environmental Medicine, Department of Public Health and Clinical Medicine, Umeå University, Sweden. [http://www.vibration.db.umu.se]. Date accessed: April 2008.

POSTURE EFFECT ON VIBRATION TRANSMISSIBILTY OF THE HAND-ARM

S. A. Adewusi,[1] S. Rakheja,[1] P. Marcotte,[2] and P.-É. Boileau[2]

[1]Concordia Centre for Advanced Vehicle Engineering (CONCAVE), Concordia University,
Montreal, Quebec, Canada
[2]Institut de recherche Robert-Sauvé en santé et en sécurité du travail (IRSST),
Montreal, Quebec, Canada

Introduction

Development of an effective model of the human hand-arm exposed to vibration requires characterizing the localized responses apart from the driving point biodynamic responses. Although the driving point responses have been widely characterized under a wide range of hand forces and postural conditions, the motions of different segments have been reported in a few studies in terms of vibration transmissibility of the finger, wrist, elbow, upper arm, and shoulder [Aatola 1989; Pyykkö et al. 1976; Reynolds and Angevine 1977]. The effect of posture, which is known to alter the biodynamic behavior of the hand-arm system, has not been adequately investigated. Moreover, the effect of push force on the transmitted vibration has not been considered. Considering the differences in experimental conditions and measurement locations used in the reported studies, extreme differences can be observed among reported transmissibility data, which are far greater than those in the reported mechanical impedance properties. The reported magnitude responses generally do not show conspicuous peaks that may be related to resonances of the hand-arm system. In this study, the vibration transmitted to different segments of the hand-arm system is characterized under different postures and hand forces.

Methods

Experiments were conducted with seven male subjects grasping a 40-mm instrumented handle while exposed to z_h-axis random vibration (a_{hw} = 5.25 m/s^2) in the 2.5–500 Hz frequency range with different combinations of grip (F_g) and push (F_p) forces. Each subject grasped the handle assuming two different hand-arm postures: 90° elbow angle with wrist in neutral position (P1) and 180° elbow angle (P2). The vibration transmitted to four different locations (wrist, elbow 1 (forearm side), elbow 2 (upper-arm side), and shoulder) of the hand-arm were measured using three-axis accelerometers attached on Velcro strips. The strips were fastened tightly near the joints to minimize the contributions due to skin artifact, while correction for skin deformations was not attempted. The transmissibility responses were obtained along the z_h- and y_h-axes at the wrist and shoulder and along the three axes at the elbow relative to the z_h-axis handle vibration using the H_1 method. Coherence of measurements was monitored to ensure reliability of the data. The transmissibility data were examined in both linear and logarithmic scales to gain a better understanding of the responses in the low- and high-frequency ranges and to identify the characteristic frequencies. ANOVA was used to study the effect of various experimental factors.

Results

Figure 1 shows the mean z_h-axis transmissibility at the selected measurement locations corresponding to the two postures considered. The results show that the z_h-axis transmissibility of the wrist is the largest at frequencies above 30 Hz and that of the shoulder is the lowest corresponding to the P1 posture. The transmissibility generally decreases from the wrist to the shoulder, although the elbow transmissibility peaks are higher near 12.5 Hz. There is attenuation of

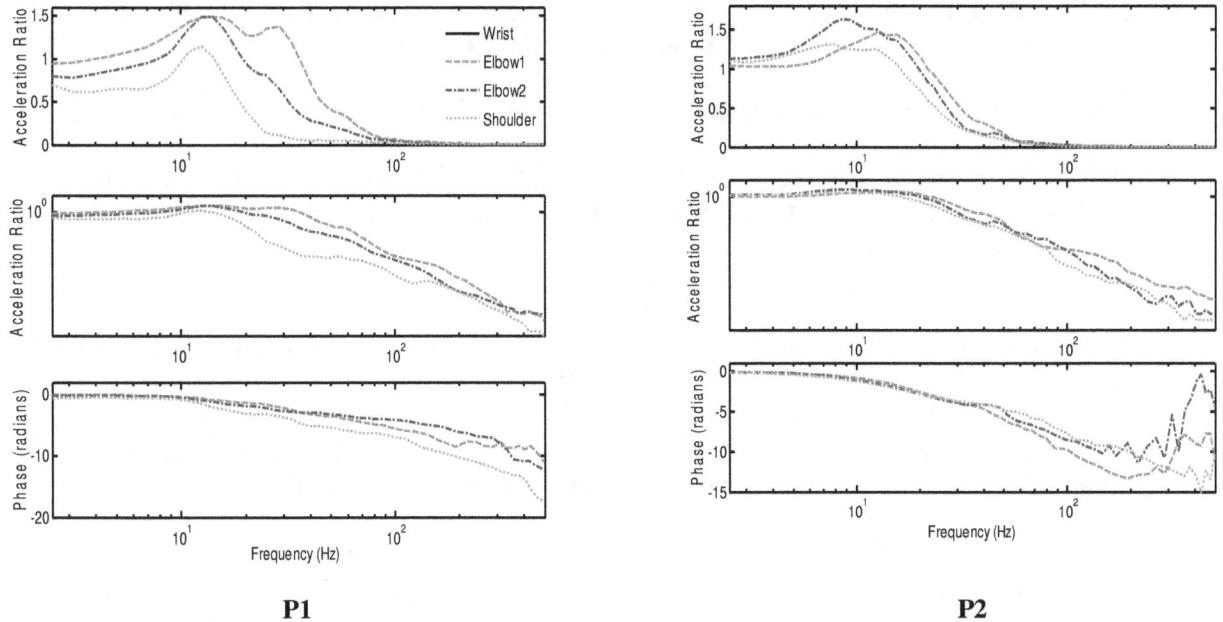

P1   **P2**

Figure 1.—Mean z_h-axis transmissibility responses (F_g = 30 N, F_p = 50 N, and a_{hw} = 5.25 m/s^2).

vibration transmitted to the upper arm for the P1 posture; thus, the elbow acts as a vibration absorber. This observation agrees with that of Pyykko et al. [1976]. The y_h-axis transmissibility of the wrist was observed to be the largest above 30 Hz, while that of elbow 2 was largest below 30 Hz. Unlike the bent-arm posture, the z_h-axis transmissibility for the extended arm (P2) posture increases from the wrist to the shoulder in the 2.5–25 Hz frequency range, while the peak magnitude at elbow 2 is the highest. Above 25 Hz, the transmissibility magnitudes decrease from the wrist to the shoulder, similar to the trend observed for the P1 posture, while the relative decrease is small. The y_h-axis transmissibility responses at the elbow 1, elbow 2, and shoulder locations for the P2 posture were greater than those for the P1 posture in the low-frequency range; the peak magnitude at the shoulder is the highest. The results thus suggest greater transmission of low-frequency vibration to the upper arm in an extended-arm posture, which may encourage rapid fatigue. The hand-arm system in an extended arm posture tends to attenuate the vibration above 25 Hz more effectively than the bent-elbow posture, which limits greater vibration to the hand and wrist. The resonant frequencies of the hand-arm system were further identified from the data for both postures, which were generally comparable with those reported in earlier studies. ANOVA results show that posture and measurement direction are significant at all frequencies, while hand forces are significant above 15 Hz.

References

Aatola S [1989]. Transmission of vibration to the wrist and comparison of frequency response function estimators. J Sound Vibration *131*(3):497–507.

Pyykkö I, Färkkilä M, Toivanen J, Korhonen O, Hyvärinen J [1976]. Transmission of vibration in the hand-arm system with special reference to changes in compression force and acceleration. Scand J Work Environ Health *2*(2):87–95.

Reynolds DD, Angevine EN [1977]. Hand-arm vibration. Part II: vibration transmission characteristics of the hand and arm. J Sound Vibration *51*(2):255–265.

A NEW BIODYNAMIC APPROACH TO ASSESS THE EFFECTIVENESS OF ANTIVIBRATION GLOVES

Ren G. Dong, Daniel E. Welcome, Thomas W. McDowell, Christopher Warren, and John Z. Wu

Engineering and Control Technology Branch, Health Effects Laboratory Division,
National Institute for Occupational Safety and Health, Morgantown, WV

Introduction

Antivibration gloves have been increasingly used to help reduce hand-transmitted vibration exposure. A critical question is how much the gloves can help. Another important question is whether the effectiveness of the gloves can be further improved. The objectives of this study were to develop a new biodynamic approach to predict the vibration transmissibility of the glove at the palm and fingers of a human hand and to analyze the mechanisms of the gloves.

Method

The proposed method is based on the measurement and modeling of the driving-point biodynamic response of the hand-arm system. First, the mechanical impedances distributed at the fingers and the palm of the hand for a bare hand and a gloved hand were measured separately using a reported method [Dong et al. 2006]. The same postures and hand forces (30-N grip and 50-N push) as those specified in ISO 10819 (1996) and a broadband random vibration were used in the measurement. Six subjects participated in the experiments using two types of gloves (glove A: air bladder; glove B: gel-filled). Next, the experimental data measured with the bare hand were used to establish a five-degree-of-freedom mechanical-equivalent model of the hand-arm system (the part with nonfilled boxes and their connections in Figure 1) using a reported method [Dong et al. 2007]. Then, based on the assumption that the biodynamic properties of the hand-arm system remain unchanged when wearing a

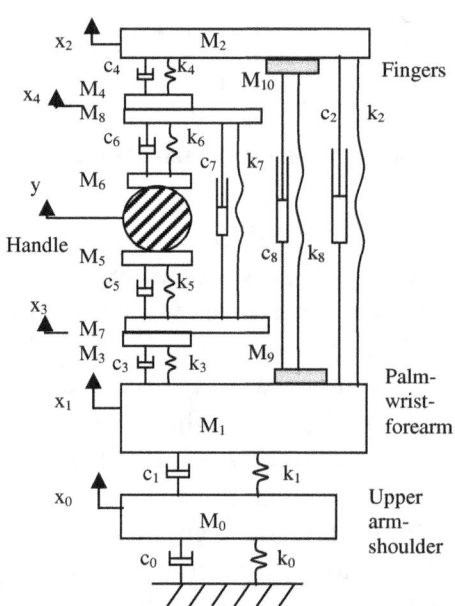

Figure 1.—A model of the gloved hand-arm system.

glove, we developed a mechanical equivalent model of the gloved hand-arm system (Figure 1) using the experimental data measured with the gloved hand. More specifically, the model parameters for the bare hand-arm system remain unchanged in the determinations of the elements for simulating the glove (the part with filled boxes and their connections in Figure 1) in the model. Finally, the glove transmissibility at the fingers is determined by taking the ratio (x_4/y) of the finger skin motion (x_4) in the gloved hand model and the handle motion (y). Similarly, the glove transmissibility at the palm is calculated by taking the ratio (x_3/y) of the palm skin motion (x_3) and the handle motion (y).

Results and Discussion

Figure 2 shows the predicted glove transmissibility values, together with those measured with the palm adapter method reported in our previous studies [Dong et al. 2003, 2004]. The basic trends of the measured and predicted transmissibility data at the palm are consistent, and their values are comparable below 200 Hz for glove A and 300 Hz for glove B. The isolation effectiveness of glove A is better than that of glove B. These observations suggest that both the proposed and the adapter method are acceptable, at least for glove-screening tests. The results also show that the gloves can provide some vibration reduction at the palm at frequencies higher than 30 Hz, but are not effective for reducing finger exposure if the vibration frequency of concern is below 250 Hz. Therefore, it is inappropriate to use the transmissibility measured at the palm to estimate the overall vibration reduction for the hand-arm system. Whereas the palm transmissibility may be used to help assess palm-wrist-arm system vibration exposure, the finger transmissibility may be applied to help assess finger vibration exposure.

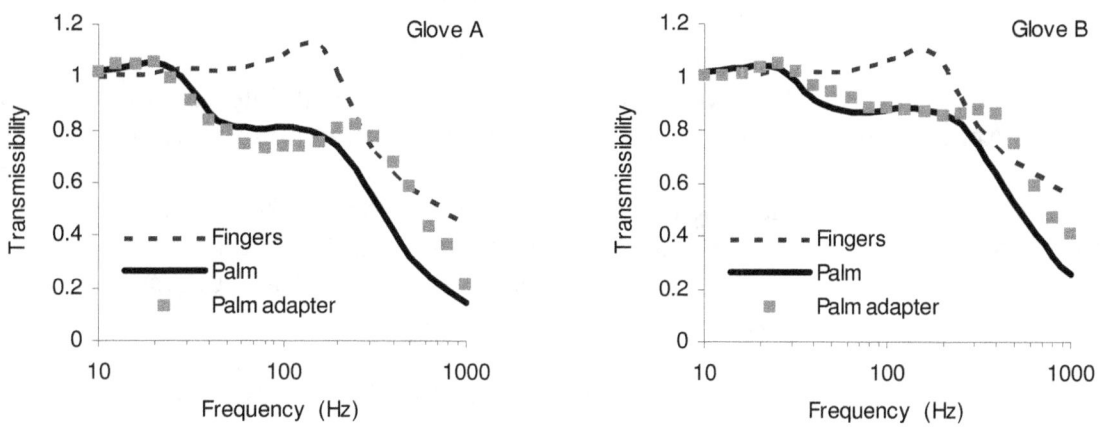

Figure 2.—Vibration transmissibility functions of an air-bladder glove (glove A) and gel-filled glove (glove B) predicted using the proposed method and measured with an adapter method [Dong et al. 2003, 2004].

In addition to the transmissibility of the gloves, the glove influences on the vibration power absorption distributed in the major substructures of the system can also be predicted using the model. Whereas it is unavoidable to impose some interference to the glove-hand-arm system in the glove test using an adapter method or an on-the-hand method, this new approach does not impose any such interference. The proposed model can be further used to identify the major factors that influence glove effectiveness and to help develop better antivibration devices.

References

Dong RG, McDowell TW, Welcome DE, Smutz WP, Schopper AW, Warren C, Wu JZ, Rakheja S [2003]. On-the-hand measurement methods for assessing effectiveness of antivibration gloves. Int J Ind Ergon *32*(4):283–298.

Dong RG, McDowell TW, Welcome DE, Barkley J, Warren C, Washington B [2004]. Effects of hand-tool coupling conditions on the isolation effectiveness of air bladder antivibration gloves. J Low Freq Noise Vibration Active Control *23*(4):231–248.

Dong RG, Welcome DE, McDowell TW, Wu JZ [2006]. Measurement of biodynamic response of human hand-arm system. J Sound Vibration *294*(4-5):807–827.

Dong RG, Dong JH, Wu JZ, Rakheja S [2007]. Modeling of biodynamic responses distributed at the fingers and the palm of the human hand-arm system. J Biomech *40*(10):2335–2340.

ISO 10819 [1996]: Mechanical vibration and shock – Hand-arm vibration – Method for the measurement and evaluation of the vibration transmissibility of gloves at the palm of the hand. Geneva, Switzerland: International Organization for Standardization. ISO 10819:1996.

Session VIII: Health Effects

Chair: Paul-Émile Boileau
Co-Chair: Christopher Jobes

Presenter	Title	Page
H. W. Paschold Ohio University	Survey of Whole-Body Vibration Awareness and Knowledge Among Safety and Health Professionals in the United States	81
I. J. H. Tiemessen University of Amsterdam	Effectiveness of an Occupational Health Intervention Program to Reduce Whole-Body Vibration Exposure	83
J. P. Holland University of Washington	The Role of Whole-Body Vibration and Repeated Shock in the Development of Neck Pain and Cervical Spine Degeneration: Results of a Systematic Literature Review	85
T. Jetzer	Lumbar Spine Pathology from Whole-Body Vibration in Truck Drivers	87
T. W. McDowell National Institute for Occupational Safety and Health (NIOSH)	Evaluating Impact Wrench Vibration Using the Method in Proposed Revisions to ISO 8662-7	89

SURVEY OF WHOLE-BODY VIBRATION AWARENESS AND KNOWLEDGE AMONG SAFETY AND HEALTH PROFESSIONALS IN THE UNITED STATES

Helmut W. Paschold
School of Health Sciences, Ohio University, Athens, OH

Introduction

Reducing occupational whole-body vibration (WBV) exposure in the United States depends on the safety and health (S&H) professional community's ability to anticipate and recognize the hazard. In the course of casual conversations with S&H practitioners, it seemed that many had not heard of WBV or had little more than a very basic understanding. An informal review indicated that most WBV literature has originated outside of the United States. In November 2006, the Ohio University Health Sciences librarian conducted an on-line search of the ISO WBV standards. It was found that only three copies were listed in North American library catalogs—two in the United States and one in Canada; none of the three is available for circulation. Currently, the Occupational Safety and Health Administration does not include or refer to WBV in its standards.

Only two previous survey studies of WBV were found. A survey of workers' self-reporting WBV exposure [Palmer et al. 2000] determined that an estimated 9 million workers in Great Britain were exposed to WBV on a weekly basis, with 4% exceeding action levels. A small survey (n=88) of WBV proficiency among construction managers and designers in the United Kingdom [Edwards and Holt 2007] revealed that S&H professionals (n=6) in the group self-reported moderate WBV knowledge, averaging 2.33 on a scale of 0 (know nothing) to 4 (know everything).

Based on casual conversations with S&H professionals relative to WBV, relatively few published articles or seminar presentations in the United States, the scarcity of available standards, and absence of mandatory federal safety regulations, an assumption was made that the topic of WBV is not very well known or understood in the United States. To explore this hypothesis, an on-line survey questionnaire about WBV knowledge was prepared and distributed to S&H professionals throughout the United States.

Methods

An on-line WBV knowledge survey for U.S. S&H professionals was designed and posted on the Web site www.SurveyMonkey.com, chosen for its ease of operation and relatively low cost of operation. The survey requested responses in three main categories:

- *Knowledge of WBV basics, health effects, standards, and monitoring* (self-rating from 1 (none) to 5 (expert) knowledge)

- *Employer size and industrial classification* (actual number of employees and industry classification according to the U.S. Census Bureau 2002 North American Industry Classification System (NAICS))

- *Characteristics of the individual professional, including training, education, experience, and certifications* (actual years of experience, university degrees, certification titles, continuing education events attended, etc.)

With gracious assistance from the American Society of Safety Engineers (ASSE), an e-mail with a link to the hosting Web site was distributed to 21,292 nonstudent ASSE members throughout the United States on May 14, 2007.

Results

A total of 2,764 persons completed the survey for a response rate of 13.0%. Of the respondents, 38.6% reported that they "haven't heard of WBV." Among the 61.4% who had heard of WBV from one or more sources, 82 persons (5%) reported no ability to define or explain WBV. Less than 1% of all respondents reported an expert-level ability to define or explain WBV, measure and quantify WBV, or knew the ISO, ANSI, BS, or ACGIH standards (all combined). Only one respondent reported expert knowledge of the British WBV Standard.

Discussion

The relatively large portion (38.6%) of the surveyed ASSE S&H professionals who "haven't heard of WBV" may be a slight underestimate of the percentage of all "unaware" S&H professionals in the United States. ASSE members tend to have higher standards of professionalism than nonmembers; some persons with no knowledge may have exited immediately without completing the survey upon reading "WBV" due to unfamiliarity; and persons tend to underreport a lack of knowledge. Further analysis of data is being performed to ascertain individual or industry-type characteristics most associated with WBV knowledge levels. The determination of the relative absence of S&H professionals' expertise in WBV suggests that anticipation, recognition, evaluation, and control of the hazard are not occurring as effectively as needed in the United States. Enhanced efforts by governmental, professional, and educational organizations are needed to increase WBV topic knowledge among S&H practitioners.

References

Edwards DJ, Holt GD [2007]. Perceptions of workplace vibration hazards among a small sample of UK construction professionals. Eng Construction Archit Manage *14*(3):261–276.

Palmer K, Griffin M, Bendall H, Pannett B, Coggon D [2000]. Prevalence and pattern of occupational exposure to whole body vibration in Great Britain: findings from a national survey. Occup Environ Med *57*(4):229–236.

EFFECTIVENESS OF AN OCCUPATIONAL HEALTH INTERVENTION PROGRAM TO REDUCE WHOLE-BODY VIBRATION EXPOSURE

Ivo J. H. Tiemessen, Carel T. J. Hulshof, and Monique H. W. Frings-Dresen

Coronel Institute for Occupational Health, Academic Medical Center, University of Amsterdam, Amsterdam, the Netherlands

Introduction

It has been estimated that 4%–7% of all drivers in some European countries, the United States, and Canada are exposed to potentially harmful whole-body vibration (WBV). Long-term occupational exposure to WBV is associated with an increased risk of low back pain and disorders of the lumbar spine [Lötters et al. 2003]. Despite this, literature on successful strategies to reduce WBV exposure in the workplace is scarce [Tiemessen et al. 2007b]. Only one study [Hulshof et al. 2006] was identified that systematically evaluated an intervention program aimed at reducing WBV exposure. Although the decrease in WBV exposure did not reach statistical significance (p=0.06), the results of this study were promising. Most of the research in this field is focused on technical prevention, although changing behavior toward WBV determining factors might be promising as well.

The goal of this study was to assess whether introducing an occupational health intervention program leads to an additional decrease in WBV exposure in work situations compared to exposure in work situations where "care as usual" is performed.

Methods

We used the ASE (attitude, social influence, self-efficacy) model [De Vries et al. 1998] as a conceptual model for the design of this intervention study. The theoretical outline of the intervention program has been described in detail elsewhere [Tiemessen et al. 2007a]. A group of 318 drivers and 9 employers were cluster randomized into 3 groups. The high-exposure (≥ 0.5 m/s^2) intervention group (group 1) received an intervention consisting of an individual health surveillance, an informational brochure, an informative oral presentation, and a report of the results of the performed measurements of the WBV magnitude. The high-exposure "care as usual" group (group 2) and the low-exposure (<0.5 m/s^2) "care as usual" group (group 3) received care as usual, which consisted only of the informational brochure and the report. At T0, the intervention program was implemented. The followup time of the study was 7 months (T1). The WBV exposure and the process variables knowledge, attitude, intended behavior, and behavior were assessed with self-administered questionnaires and by field vibration measurements according to ISO 2631-1. Data were analyzed using SPSS (version 13.0). To describe the effects on WBV exposure and on the process variables due to implementation of the intervention program, we used a linear mixed model. Adjustments for the cluster randomization to account for both the effects on company and individual level were made. The difference in WBV exposure at T0 and T1 was entered as a continuous outcome variable. Because of nonnormality of the data, we assigned ranking to this variable and used a nonparametric model.

Results

The response of drivers was 75% (n=240) at T0 and 65% (n=206) at T1. The response of the employers/supervisors of the companies was 100% (n=9) at T0 and 89% (n=8) at T1. Of the 240 drivers at T0, 36% (n=86) seemed to exceed the EU action value for daily exposure of 0.5 m/s^2 during their work activities.

Baseline characteristics showed no relevant differences at T0 between the groups. Mixed model analysis revealed no significant difference in WBV exposure between groups 1 and 2. At T1, however, we found a decrease in WBV for 23% of the drivers in group 1 below the 0.5 m/s^2 EU action value. Mixed model analysis also revealed no significant difference in knowledge, attitude, intended behavior, and actual behavior (see Table 1). Compliance in group 1 to the agreed changes during the health surveillance was low; only 19 drivers were compliant for more than 50%.

Table 1.—Gain score (T1 scores-T0 scores) for the three defined groups and the employers

	Group 1 (n=65)	Group 2 (n=61)	Group 3 (n=192)
Driver			
WBV - gain	−0.04 ± 0.21	0.00 ± 0.13	0.11 ± 0.14
Knowledge - gain	0.25 ± 2.75	0.66 ± 1.63	0.43 ± 2.13
Attitude - gain	−0.97 ± 5.85	−0.31 ± 6.18	0.02 ± 5.46
Intended behavior	2.24 ± 1.69	2.67 ± 2.06	2.43 ± 2.18
Behavior - gain	0.66 ± 5.26	−0.23 ± 5.07	−1.06 ± 4.94
Employer/supervisor	Intervention group	"Care as usual" group	
Knowledge - gain	0.2 ± 1.10	−1.33 ± 3.79	
WBV - policy	1.4 ± 3.51	4.0 ± 4.36	

The implementation failed to increase the knowledge of the employers both in the intervention and the "care as usual" groups, but a tendency for an increase in company policy was found (see Table 1).

Discussion

Implementation of the presented intervention program was not effective in reducing the average WBV exposure on group level significantly. However, for a substantial portion of the drivers in the intervention group, the daily vibration exposure was reduced to a level below the EU action value. Besides aiming at technical preventive measures, future research should also focus on the role of the individual driver in preventing adverse effects of WBV exposure.

References

De Vries H, Mudde AN, Dijkstra A, Willemsen MC [1998]. Differential beliefs, perceived social influences, and self-efficacy expectations among smokers in various motivational phases. Prev Med 27(5):681–689.

Hulshof CTJ, Verbeek JHAM, Braam ITJ, Bovenzi M, van Dijk FJH [2006]. Evaluation of an occupational health intervention programme on whole-body vibration in forklift truck drivers: a controlled trial. Occup Environ Med 63:461–468.

Lötters F, Burdorf A, Kuiper J, Miedema H [2003]. Model for the work-relatedness of low-back pain. Scand J Work Environ Health 29:431–40.

Tiemessen IJH, Hulshof CTJ, Frings-Dresen MHW [2007a]. The development of an intervention programme to reduce whole-body vibration exposure at work induced by a change in behaviour: a study protocol. BioMed Cent Public Health, Nov 15;7:329.

Tiemessen IJH, Hulshof CTJ, Frings-Dresen MHW [2007b]. An overview of strategies to reduce whole-body vibration exposure on drivers: a systematic review. Int J Ind Ergon 37(3): 245–256.

THE ROLE OF WHOLE-BODY VIBRATION AND REPEATED SHOCK IN THE DEVELOPMENT OF NECK PAIN AND CERVICAL SPINE DEGENERATION: RESULTS OF A SYSTEMATIC LITERATURE REVIEW

John P. Holland,[1] Carole L. Holland,[2] and John Webster[3]

[1]Principal, Holland and Associates, Seattle, WA
Clinical Faculty, Department of Environmental and Occupational Health Sciences,
University of Washington, Seattle, WA

[2]Principal, Holland and Associates, Seattle, WA
Affiliate Faculty, Seattle University, Seattle, WA

[3]Consultant, San Diego, CA
Retired, Chief of Orthopaedic Surgery, San Diego Naval Medical Center, San Diego, CA

Background

Whole-body vibration (WBV) has been proposed as a risk factor for the development of neck pain secondary to degeneration of spinal structures. We conducted a systematic review of the scientific literature to evaluate the strength of the evidence for an association between WBV and neck pain or cervical spine degeneration.

Methods

A comprehensive search was conducted of bibliographic databases through August 2007. Two investigators independently reviewed all abstracts to identify relevant articles. Full-text versions of potentially relevant articles were systematically evaluated. All articles containing original data were graded for quality of study design and precision of exposure and outcome classification. Due to heterogeneity in study designs, exposure classification, and outcome classification, meta-analysis was not attempted.

Adverse health outcomes were stratified into five categories: (1) self-reported symptoms of neck pain, (2) symptoms of neck pain and consultation with a health care provider, (3) symptoms of neck pain and evidence of cervical spine degeneration on imaging, (4) injury claim or leave of absence for neck pain derived from administrative databases, and (5) symptoms of neck pain, evidence of cervical spine degeneration on imaging, and positive electrodiagnostic evidence for radiculopathy.

Exposure estimates were stratified into four categories: (1) occupational title, (2) self-reported exposure to WBV or history of occupational driving, (3) quantification of time exposed to WBV on a specific job, and (4) measurement of WBV according to national or international consensus standards. No article was excluded based on classification of disease or exposure estimates. However, greater weight was assigned to articles with objective or quantitative measures of exposure and clinical outcome.

A qualitative synthesis of study results was done using generally accepted guidelines for the determination of causality, including (1) strength of the association, (2) consistency of the association, (3) specificity of the association, (4) temporality of the association, (5) biological plausibility of the association, (6) coherence of the association, (7) experimental evidence for the association, and (8) dose-response relationship.

Results

The bibliographic search identified 269 relevant citations, and 60 articles were selected for systematic review. Most of the epidemiological studies evaluated occupational groups, including occupational drivers, heavy equipment operators, aircraft pilots, and industrial workers. Generally, classification of adverse health consequences was based on self-reported symptoms of neck pain derived from questionnaires. Indices of severity, typically consultation with any health care provider for neck pain or lost time from employment, were derived from questionnaires. Exposure to WBV was typically classified based on occupational title or self-reported exposure. Quantitative measurements of WBV exposure (i.e., frequency, intensity, duration, and temporal patterns of exposure) were typically not reported.

Conclusions

Presently, there is insufficient evidence of a causal association between WBV and neck pain or cervical spine degeneration. Generally, study designs were inadequate, WBV exposure was not quantified, outcome definitions lacked specificity, and potential confounders were not rigorously examined.

LUMBAR SPINE PATHOLOGY FROM WHOLE-BODY VIBRATION IN TRUCK DRIVERS

Thomas Jetzer
Occupational Medicine Consultants, Minneapolis, MN

Introduction

It has long been postulated that the lumbar spines of drivers of industrial vehicles are adversely affected by whole-body vibration (WBV) transmitted through their seats. The literature contains many studies that imply that WBV from driving industrial vehicles can cause various levels of pathology to the lumbar spine [Andrusaitis et al. 2006; Camerino et al. 1997]. However, these studies have had difficulty isolating the effects of WBV from other confounding factors that can affect the integrity of the lumbar spine, including the effect of aging, excess weight, posture, associated lifting tasks, and genetic-related variability [Seidel 2005; Lings and Leboeuf-Yde 1998]. Analysis of the effect of WBV on the lumbar spine can be addressed from a number of perspectives, including the mechanical aspects often characterized by spinal modeling [Matsumoto and Griffin 2001; Seidel et al. 2001] due to difficulty of such measurement in the in vivo environment, cellular damage analysis either from the biochemical perspective or by way of various imaging methods, or by epidemiological analysis [Drerup et al. 1999; Hirano et al. 1988]. While many studies have implied that WBV is a significant cause of lumbar spine pathology, this opinion is valid only if it is supported by analysis of data from the work environment.

The purpose of this study was to assess the incidence of low back pathology in drivers from various industries. The industries were chosen to show variation in the drivers' tasks with regard to lifting and physical activities. Airline pilots were used as controls to negate the effect of posture and seating endurance. The subjects were selected randomly from the clients of an occupational clinic, with repeated visits of the same individuals to ensure consistency in medical history and physical findings. Data were analyzed to determine if there is any correlation between driving, job activities, or physical and medical information to support the implication that vibration associated with driving activities can cause lumbar spine pathology.

Method

Chart analyses of 1,697 drivers from various industries, including over-the-road truckers, trash collector drivers, utility linemen, movers/drivers, and school bus drivers, were compared to those of a similar group of airline pilots. The points of comparison included age, body mass index (BMI), work history, and medical history. The aforementioned industries were selected because of the differences in physical activity required by the jobs along with significant and similar amounts of driving activity. A comparison was made between these groups of drivers to determine if there was a significant difference in any of the parameters.

Results

Analysis of the comparison parameters indicated that the strongest association with lumbar spine pathology was weight, BMI, and job activity. Bus drivers and the control group of pilots had similar profiles of lumbar pathology. The level of pathology was most associated with nonvibration related to age, work activities, and BMI in all groups. There was very little pathology that was not explainable by a review of medical histories independent of any driving activities.

Discussion

The results of this study indicate that lumbar spine pathology can be explained by confounding factors of jobs, work history, age, and BMI characteristics independent of driving activities. In other words, BMI, age, and work history can explain the occurrence of spinal pathology in drivers as opposed to any vibration to which they may be exposed in the job. While there is evidence that individuals with significant spinal preexisting pathology may have poor tolerance from WBV due to vibration resonance with abnormal spinal and neurological architecture [Vanharanta et al. 1998; Yrjämä et al. 1996], the evidence does not support that this pathology is caused by vibration in industrial driving addressed in this study. With the advent of new seat technologies that are widely deployed throughout the trucking industry, there is little basis to state that workers are exposed to enough WBV to result in significant pathology.

References

Andrusaitis SF, Oliveira RP, Barros Filho TE [2006]. Study of the prevalence and risk factors for low back pain in truck drivers in the state of São Paulo, Brazil. Clinics *61*(6):503–510.

Camerino D, Molteni G, Volponi R, Simionato B, Dondè I [1997]. Public transportation driving and disorders of their vertebral spine: subjective evaluation of the risks (in Italian). Med Lav *88*(5):382–395.

Drerup B, Granitzka M, Assheuer J, Zerlett G [1999]. Assessment of disc injury in subjects exposed to long-term whole-body vibration. Eur Spine J *8*(6):458–467.

Hirano N, Tsuji H, Ohshima H, Kitano S, Itoh T, Sano A [1988]. Analysis of rabbit intervertebral disc physiology based on water metabolism. II. Changes in normal intervertebral discs under axial vibratory load. Spine *13*(11):1297–1302.

Lings S, Leboeuf-Yde C [1998]. Whole body vibrations and low back pain (in Danish). Ugeskr Laeger *160*(29):4298–4301.

Matsumoto Y, Griffin MJ [2001]. Modelling the dynamic mechanisms associated with the principal resonance of the seated human body. Clin Biomech (Bristol, Avon) *16*(Suppl 1):S31–44.

Seidel H [2005]. On the relationship between whole-body vibration exposure and spinal health risk. Ind Health *43*(3):361–377.

Seidel H, Bluthner R, Hinz B [2001]. Application of finite-element models to predict forces acting on the lumbar spine during whole-body vibration. Clin Biomech (Bristol, Avon) *16*(Suppl 1):S57–63.

Vanharanta H, Ohnmeiss DD, Aprill CN [1998]. Vibration pain provocation can improve the specificity of MRI in the diagnosis of symptomatic lumbar disc rupture. Clin J Pain *14*(3):239–247.

Yrjämä M, Tervonen O, Vanharanta H [1996]. Ultrasonic imaging of lumbar discs combined with vibration pain provocation compared with discography in the diagnosis of internal anular fissures of the lumbar spine. Spine *21*(5):571–575.

EVALUATING IMPACT WRENCH VIBRATION USING THE METHOD IN PROPOSED REVISIONS TO ISO 8662-7

T. W. McDowell, R. G. Dong, X. Xu, D. E. Welcome, and C. Warren

Health Effects Laboratory Division, National Institute for Occupational Safety and Health, Morgantown, WV

Introduction

ISO 8662-7, the international standard for evaluating vibration emissions from impact wrenches, is undergoing a systematic review and revision. The revisions are designed to more closely simulate actual vibration in the workplace, ensure that the end results will fall into the top quartile of what is actually experienced by a variety of workers in different work situations, produce a vibration result that could be easily translated into risk assessment analysis, and be able to be applied to a cross-section of all threaded fastener tools [ISO 2006]. While the current standard calls for single-axis vibration measurements to be made while the tool acts against a braking device, the revised procedure specifies that triaxial vibration measurements be made over the course of a series of 30-sec trials that involve the seating of 10 nuts onto plate-mounted studs or bolts.

In order to provide information toward the improvement of the impact wrench testing method, NIOSH constructed the proposed test rig and conducted impact wrench evaluations using the revised procedure. The objectives of this study were to compare methods for quantifying impact wrench vibrations and perform a general evaluation of the proposed revisions to ISO 8662-7.

Methods

Six male experienced impact wrench operators used 15 impact wrenches in the simulated work task. Three samples each of four pneumatic models (A through D) and one battery-powered model (model E) were used. The test rig consisted of two steel plates vertically mounted on a concrete block. The plates featured channels and holes to accommodate 10 steel bolts arranged in 2 evenly spaced rows. Each bolt was fitted with a nut, two Belleville washers, and a matching flat washer. Triaxial vibration measurements were made over the course of a series of 30-sec trials that involved the seating of 10 nuts onto the bolts. Each operator completed five 10-nut trials with each of the 15 tools for a total of 75 trials in a test session.

For the four pneumatic tools, vibration was measured at the tool handle and at the front portion of the tool housing. For the smaller battery-powered tool, only handle vibration was measured. Vibration data were collected for each one-third octave band with center frequencies from 6.3 to 1250 Hz.

Results and Discussion

As expected, the within-tool acceleration values varied less than the within-operator values. ANOVAs for ISO-weighted (ISO 5349-1) and unweighted acceleration each revealed that both tool and operator are significant factors, as well as their interaction ($p < 0.001$). However, as indicated in Table 1, an evaluator rank-ordering the tools by acceleration would draw different conclusions depending on whether or not ISO weighting was applied. Thus, it may be best to report both acceleration values.

As shown in Figure 1, there is a strong correlation between ISO-weighted acceleration measured at the tool handle and that measured on the motor housing ($R^2 = 0.875$, $p < 0.001$). This relationship holds true for unweighted acceleration. Thus, for tool-screening purposes, it may not be important to measure vibration at multiple tool locations.

Table 1.—Vibration measured at the tool handle for each subject and tool type

ISO-weighted acceleration (m/s^2)

Tool Operator	Model B Mean	SD	Model A Mean	SD	Model E Mean	SD	Model D Mean	SD	Model C Mean	SD
3	2.7	0.2	5.5	0.5	7.1	0.3	6.5	0.3	8.0	1.0
2	2.7	0.1	6.0	0.4	6.6	0.6	6.8	0.6	7.9	0.7
5	3.2	0.3	6.3	0.3	6.4	0.8	7.7	0.1	8.8	0.8
1	2.6	0.3	6.4	0.8	6.5	0.5	8.2	1.4	8.9	2.5
6	3.2	0.5	7.1	0.7	7.3	0.8	7.7	0.7	10.3	1.5
4	3.0	0.2	7.6	0.5	8.3	0.8	9.9	0.8	9.7	0.7
Average	2.9	0.3	6.5	0.5	7.0	0.6	7.8	0.6	8.9	1.2

Unweighted acceleration (m/s^2)

Tool Operator	Model B Mean	SD	Model A Mean	SD	Model E Mean	SD	Model D Mean	SD	Model C Mean	SD
3	27.0	3.3	100.6	9.3	202.4	3.3	173.5	10.0	129.2	6.8
2	26.4	1.5	114.0	10.4	153.0	16.4	209.6	10.8	134.1	17.6
5	35.3	1.9	119.0	5.8	147.3	21.1	209.0	5.9	155.5	15.7
1	30.1	6.4	109.0	20.4	145.6	14.9	219.0	34.4	161.6	37.8
6	30.8	1.6	138.5	3.2	161.1	15.4	228.7	11.2	210.4	32.3
4	29.2	1.9	130.5	3.8	191.1	26.6	251.6	20.9	192.2	18.9
Average	29.8	2.8	118.6	8.8	166.7	16.3	215.2	15.5	163.8	21.5

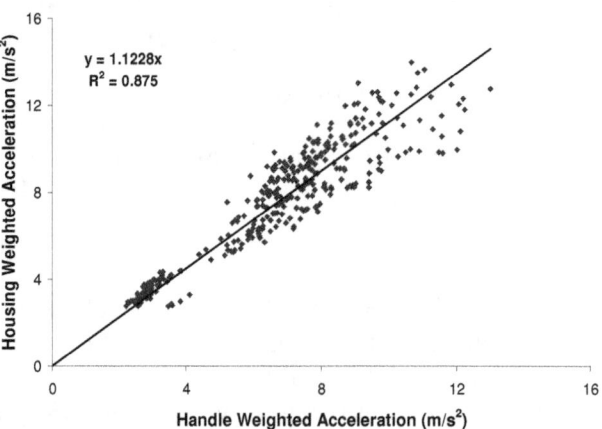

Figure 1.—Handle versus housing vibration.

Both the current and proposed impact wrench assessment standards call for testing by three tool operators. However, a power analysis [Montgomery 2001] revealed that three tool operators may not be enough for reliable tool screening. Based on the acceleration variance observed in this study, at least five tool operators would be required to reliably rank-order all five tool models using this revised procedure ($\alpha = 0.05$, $\beta = 0.20$, $p < 0.05$).

It is generally recognized that work posture and applied hand forces can significantly affect hand-transmitted vibration exposure [Griffin 1997]. In light of these observations, it may be useful to modify the impact wrench test procedure in an effort to reduce variability in the results. For example, to better control work posture, the operator could stand on a platform with adjustable height. Thus, elbow and shoulder angles could be more standardized. To help evaluate the influence of applied forces, a force plate could be added to the platform to measure ground reaction forces.

While this test setup may not provide reasonable simulations for all work tasks involving impact wrenches, the apparatus and test procedure seem to generate appropriate data for tool screening for a wide variety of impact wrench models and many common workplace operations.

References

Griffin MJ [1997]. Measurement, evaluation, and assessment of occupational exposures to hand-transmitted vibration. Occup Environ Med *54*(2):73–89.

ISO [1997]. Hand-held portable power tools – Measurement of vibrations at the handle – Part 7: Wrenches, screwdrivers and nut runners with impact, impulse or ratchet action. Geneva, Switzerland: International Organization for Standardization. ISO 8662-7:1997.

ISO [2001]. Mechanical vibration – Measurement and evaluation of human exposure to hand-transmitted vibration – Part 1: General requirements. Geneva, Switzerland: International Organization for Standardization. ISO 5349-1:2001.

ISO [2006]. Round robin test setup and procedure for proposed changes to ISO 8662-7. ISO/TC 118/SC 3/WG 3 – Vibrations in hand-held tools – Ad hoc group for wrenches.

Montgomery DC [2001]. Design and analysis of experiments. 5th ed. New York: John Wiley & Sons, pp. 107–109.

Session IX: Epidemiology and Standards II

Chairman: Danny Riley
Co-Chair: Sandya Govindaraju

Presenter	Title	Page
D. Riley Medical College of Wisconsin	Effect of an FMRI-Compatible Piezoelectric Vibration Actuator on Rat Tail Skin Perfusion	92
M. A. Loffredo Medical College of Wisconsin	Vibration-induced Nerve Injury and Recovery in a Rat Tail Model	94
S. Govindaraju Medical College of Wisconsin	Effects of Vibration Exposure on Arterioles in Rat Tail Nerves	96
K. Krajnak National Institute for Occupational Safety and Health (NIOSH)	Reduced Sensory Nerve Thresholds May Be Indicative Of Nerve Injury After Repeated Exposure To Vibration	98
A. J. Brammer University of Connecticut Health Center and National Research Council (Canada)	Influence of Change in Work Practices on Vibrotactile Perception: Preliminary Analysis of a Prospective Study of Forestry Workers	100
Y. Ye University of Southampton	Individual Variability in Changes in Finger Blood Flow Induced by Hand-Transmitted Vibration	102
S. Živanović * University of Sheffield	Vibration Exposure in Offices: A Case Study	104
J. Starkman Midwestern University	The Determination of Osteogenic Frequencies Related to the Osseointegration of Implanted Devices	106
T. Lutz Wacker Neuson SE	Vibration Guidelines: Overregulation in Europe Puts Unjustified Burden on the Construction Industry	108

* Author unable to attend conference.

EFFECT OF AN fMRI-COMPATIBLE PIEZOELECTRIC VIBRATION ACTUATOR ON RAT TAIL SKIN PERFUSION

Danny Riley, Sandya Govindaraju, Chris Zahm, and Keisha Rogers
Department of Cell Biology, Neurobiology and Anatomy, Medical College of Wisconsin,
Milwaukee, WI

Introduction

Vibration of the extremities causes immediate vasoconstriction. This is well documented in both human and animal models [Bovenzi et al. 1995; Okada 1986]. In the rat tail vibration model, tail skin perfusion fell 37% when exposed to 60 Hz vibration for 5 min [Curry et al. 2005]. The vasoconstrictive response to vibration has been attributed to activation of the somatosympathetic pathway in response to stimulation of pacinian vibroreceptors in the skin [Ekenvall and Lindblad 1986; Olsen 1990; Stoyneva et al. 2003]. The somatosympathetic response is hypothesized to be mediated within the rostroventrolateral medulla in the central nervous system, but direct evidence is lacking [Ganong 1999].

Functional magnetic resonance imaging (fMRI) detects elevated neuronal activity by the BOLD signal from radiofrequency impulses perturbed by increased blood flow in a high-intensity magnetic field environment. Evidence for or against the involvement of the somatosympathetic system in vibration-induced vasoconstriction can be demonstrated by fMRI of the rostroventrolateral medulla during finger vibration exposure.

Traditional vibration motors are made of ferrous materials and cannot be used for fMRI because they are strongly attracted to the MRI magnet. To address this problem, an MRI-compatible vibration actuator was developed by Dynamic Structure and Materials (DSM), Franklin, TN (Figure 1). Using the rat tail vibration model, the present study sought to determine whether the piezoelectric vibration actuator produces a decrease in skin perfusion similar to our B&K vibration motor (model 4809, Bruel & Kjaer North America, Norcross, GA) [Curry et al. 2005].

Figure 1.—The MRI-compatible flextensional piezoelectric actuator is constructed of titanium and sheets of piezoelectric material (black arrow) that bend in response to applied voltage to cause vertical translation of the lever arm platform.

Methods

Sprague Dawley male rats (n=10, 250–300 g) were restrained in tube cages with their tails, weighing 6.9 ± 0.1 g, taped to the vibrating platform of the piezoelectric actuator. Blood perfusion to the tail skin was measured at the level of tail segment 7 with a laser Doppler flowmeter (Transonic Systems BLF21 monitor, HL-I1047 probe, Ithaca, NY). For each rat, the baseline perfusion level was monitored for 5 min before the tail was exposed 5 min to 30 Hz vibration of 2 mm peak-to-peak displacement, a calculated 25.1 m/sec² rms. Skin perfusion was monitored for 30 min postvibration.

Results

Tail skin perfusion was decreased by 5 min of vibration and returned to the previbration baseline level quickly after vibration ceased (Figure 2).

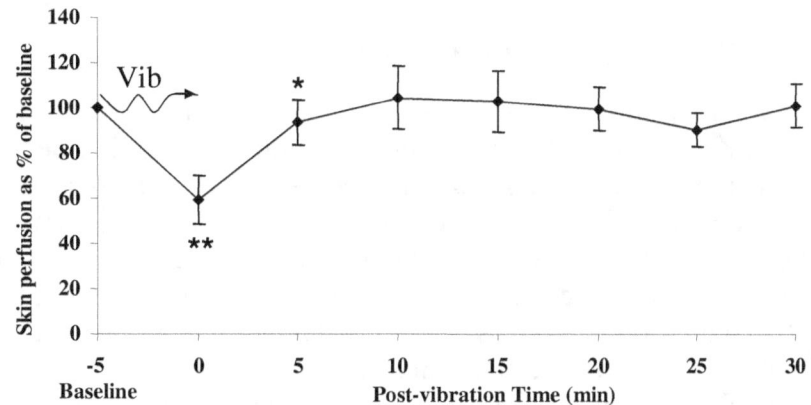

Figure 2.—Tail skin perfusion is plotted as the percentage of the pre-vibration baseline perfusion. The wavy arrow denotes the 5-min period of vibration. Tail perfusion was decreased after 5 min vibration. Perfusion returned to baseline within 5 min of stopping vibration and remained stable. Values are presented as mean ± SEM. (Double asterisk = significantly different from baseline at p<0.005. Single asterisk = significantly different from time 0 at p<0.05).

Conclusions

The DSM piezoelectric actuator is able to reduce tail perfusion in a manner similar to the B&K 4809 motor [Curry et al. 2005]. The actuator seems to be adequate for finger vibration in the fMRI environment.

References

Bovenzi M, Griffin MJ, Ruffell CM [1995]. Vascular responses to acute vibration in the fingers of normal subjects. Cent Eur J Public Health 3 Suppl:15–18.

Curry BD, Govindaraju SR, Bain JL, Zhang LL, Yan JG, Matloub HS, Riley DA [2005]. Nifedipine pretreatment reduces vibration-induced vascular damage. Muscle Nerve 32(5):639–46.

Ekenvall L, Lindblad LE [1986]. Is vibration white finger a primary sympathetic nerve injury? Br J Ind Med 43(10):702–706.

Ganong WF [1999]. Review of medical physiology. Appleton & Lange.

Okada A [1986]. Physiological response of the rat to different vibration frequencies. Scand J Work Environ Health 12(4 Spec No):362–364.

Olsen N [1990]. Hyperreactivity of the central sympathetic nervous system in vibration-induced white finger. Kurume Med J 37 Suppl: S109–116.

Stoyneva Z, Lyapina M, Tzvetkov D, Vodenicharov E [2003]. Current pathophysiological views on vibration-induced Raynaud's phenomenon. Cardiovasc Res 57(3):615–624.

VIBRATION-INDUCED NERVE INJURY AND RECOVERY IN A RAT TAIL MODEL

Michael A. Loffredo,[1] Dennis Kao,[1] Ji-Geng Yan,[1] Lin-Ling Zhang,[1] Danny A. Riley,[2] and Hani S. Matloub[1]

[1]Department of Plastic Surgery, Medical College of Wisconsin, Milwaukee, WI
[2]Department of Cell Biology, Neurobiology, and Anatomy, Medical College of Wisconsin, Milwaukee, WI

Introduction

Hand-arm vibration syndrome (HAVS), a condition consisting of neurological and vascular dysfunction, is caused by exposure to hand-operated vibration tools. The development of HAVS is dose-dependent, with patients clinically presenting with a secondary Raynaud's phenomenon (vasoconstriction) as well as sensorineural disturbances. Eight million to ten million U.S. workers are exposed to vibration tools daily [Curry et al. 2002]. Via a rat tail model, we have demonstrated vibration-induced disruption of retrograde axoplasmic transport after only two 5-hr long periods of vibration, as well as structural changes to endothelial cells after a single 4-hr exposure to vibration [Curry et al. 2002; Yan et al. 2005]. The present study was conducted to investigate the effect of vibration exposure duration on the disruption of nerve conduction velocity (NCV) and the ability to recover nerve function.

Methods

Forty-eight male Sprague Dawley rats were randomly assigned to one of six groups, with four of eight rats in each group serving as a nonvibrated control. The controls were processed and handled identically with the vibrated animals, but their tails were taped to a nonvibrating platform. The vibrated rat tails were exposed to linear vertical oscillations of 60 Hz and 5-g (49 m/s^2) acceleration. The daily duration of vibration was constant at 4 hr/day, and the total exposure was either 7 or 14 days.

Group	No. of days of vibration	No. of days of recovery
A	14	60
B	7	60
C	14	30
D	7	30
E	7	0
F	14	0

Rats were then deeply anesthetized with Nembutal. Nerve injury was evaluated electrophysiologically by recording the NCV in the ventral lateral nerve trunk of the tail on the final day of recovery, after which the tail segments C5/C6 were harvested for nerve, artery, skin, and muscle samples for further analysis. The rats were then euthanized. Animal treatment and surgical interventions were approved by the Medical College of Wisconsin Animal Care Committee and complied with the Laboratory Animal Welfare Act.

Results

The mean NCV of the control subsets within each group were not different statistically. Therefore, we combined all 24 control rats into one group (total control), and the mean NCV of this group was 3.95 cm/ms ± 0.21. The mean NCV of the vibrated rats is shown in Figure 1.

When compared to the total control group, groups A, C, D, E, and F demonstrated a statistically significant difference, with groups A, C, E, and F having a P value < 0.001 and group D having a P value < 0.05. Group B was not statistically different from the total control group.

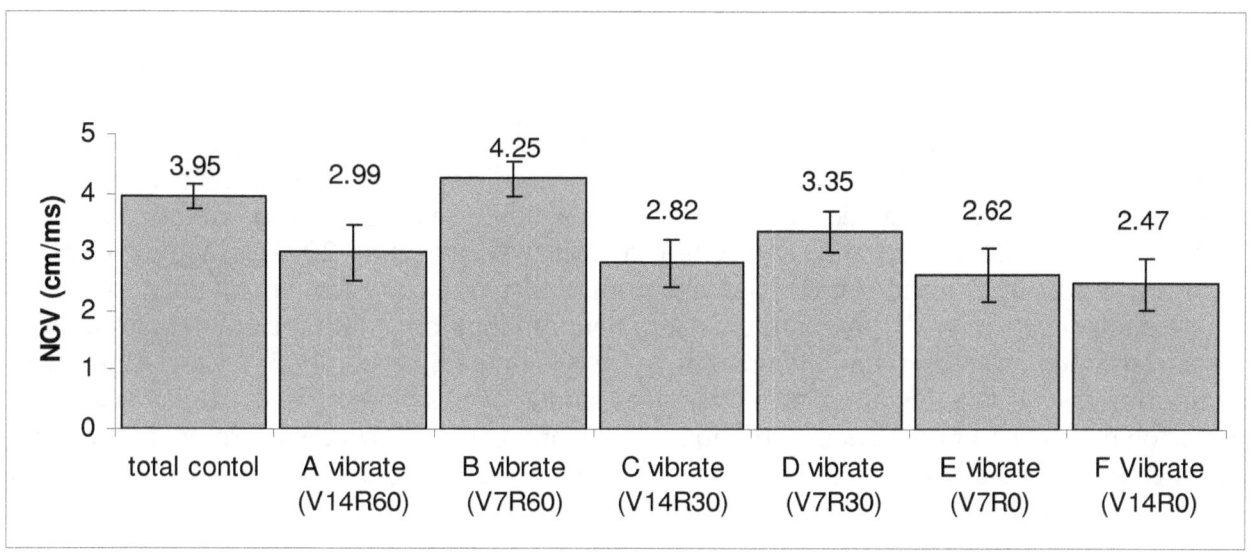

Figure 1.—NCV in vibrated rat tails. V14R60 – vibrate 14 days, recover 60 days; V7R60 – vibrate 7 days, recover 60 days; V14R30 – vibrate 14 days, recover 30 days; V7R30 – vibrate 7 days, recover 30 days; V7R0 – vibrate 7 days, recover 0 days; V14R0 – vibrate 14 days, recover 0 days. Error bars are SEM.

Discussion

Intense and long duration, hand-transmitted vibration causes a debilitating neurological, vascular, and musculoskeletal syndrome known as HAVS. The rat tail model is a valuable tool in studying the effects of vibration injury in HAVS because the tail has structural similarities to the human hand and digits [Curry et al. 2002]. Prior studies have shown that the NCV of a rat tail nerve is significantly reduced after a single 4-hr episode of vibration at 60 Hz and acceleration of 50 m/s^2 [Okada 1986]. The present study demonstrates the effect on NCV after prolonged vibration and, more importantly, displays the ability of the rat tail nerve to recover from 7-day exposure but not 14-day exposure. Vibration for 14 days causes a decrease in NCV that does not return to normal after a recovery period of 60 days. Vibration for 7 days also causes a significant decrease in NCV, but the injury, although still significant, shows some recovery 30 days post-vibration (group D) and complete recovery 60 days following vibration. Our findings suggest that 14-day vibration may produce irreversible nerve conduction damage or may require more than 60 days to recover. Thus, recovery is a slower process than injury.

References

Curry BD, Bain JLW, Yan J-G, Zhang L-L, Yamaguchi M, Matloub HS, Riley DA [2002]. Vibration injury damages arterial endothelial cells. Muscle Nerve 25(4):527–534.

Okada A [1986]. Physiological response of the rat to different vibration frequencies. Scand J Work Environ Health 12(4 Spec No):362–364.

Yan J-G, Matloub HS, Sanger JR, Zhang L-L, Riley DA [2005]. Vibration-induced disruption of retrograde axoplasmic transport in peripheral nerve. Muscle Nerve 32(4):521–526.

EFFECTS OF VIBRATION EXPOSURE ON ARTERIOLES IN RAT TAIL NERVES

Sandya Govindaraju, James Bain, and Danny Riley
Department of Cell Biology, Neurobiology, and Anatomy, Medical College of Wisconsin,
Milwaukee, WI

Introduction

Vasospastic blanching of the fingers and peripheral neuropathy are major complications of hand-arm vibration syndrome, an occupational disorder in workers using handheld power tools. Our rat tail vibration model was developed to simulate hand-transmitted vibration and investigate the cellular mechanisms of vibration injury [Curry et al. 2002]. A single 4-hr exposure to 30, 60, 120, and 800 Hz causes arteriolar dilatation and edema in rat tail nerves [Govindaraju et al., in press]. A continuous 4-hr bout of vibration at 60 Hz causes cycles of vasoconstriction and relaxation in the rat tail artery [Govindaraju et al. 2007b]. Laser Doppler monitoring reveals that the rat tail skin undergoes similar cycles of decreased and increased perfusion during a 4-hr continuous vibration [Govindaraju et al. 2007a]. The present study was designed to test if the arterioles in tail nerves undergo cyclic changes in lumen size during a 4-hr exposure to continuous vibration at 60 Hz, 49 m/s^2 (rms) acceleration.

Methods

Awake rats were divided into five groups of seven to eight animals per group and restrained on a nonvibrating platform with their tails taped to a vibrating stage. Four groups were vibrated with a frequency of 60 Hz, peak-to-peak amplitude of 0.98 mm, and an acceleration of 49 m/s^2 (rms) for 5 min, 1 hr, 2 hr, and 4 hr. One group served as a sham control and was treated similarly for 4 hr but not vibrated. Following vibration, the tail segment C7 was removed and immersion-fixed in 4% glutaraldehyde, 2% paraformaldehyde in cacodylate buffer (pH 7.4). Nerves were postfixed in 1.3% osmium tetroxide and embedded in epoxy resin for semithin (0.5 μm) sectioning. Version 1.28v ImageJ software was used to measure lumen diameter in toluidine blue-stained semithin sections.

Results

Immediately following 5 min vibration, the nerve arterioles were significantly ($p < 0.05$) dilated compared to the sham group (Figures 1–2). The 1-hr vibration group had the largest lumens. After 2 hr vibration, lumens were comparable to the sham vibration group. At the end of the 4-hr exposure, arterioles were again dilated.

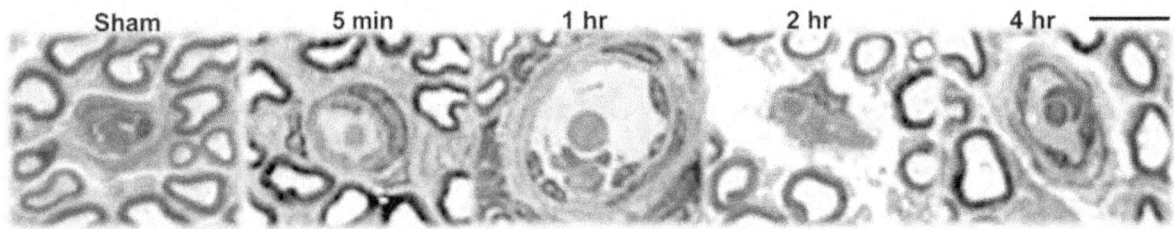

Figure 1.—Semithin cross-sections of tail nerves from rats exposed to vibration. Compared to sham, arteriole lumens are larger at 5 min, 1 hr, and 4 hr. Lumen size at 2 hr is similar to sham. Bar equals 10 μm.

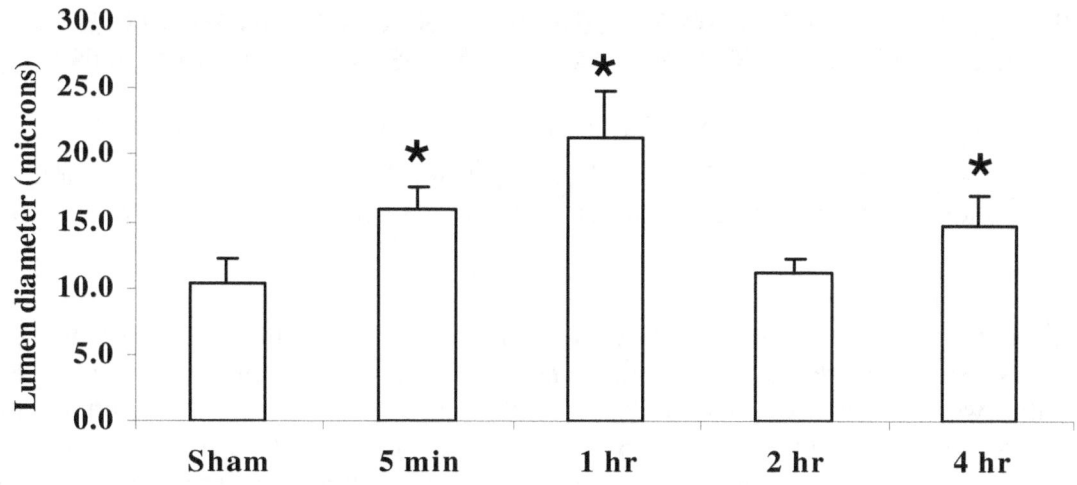

Figure 2.—Comparison of lumen diameters of nerve arterioles. Lumen diameters were significantly greater than sham at 5 min, 1 hr, and 4 hr (asterisk indicates $p < 0.05$).

Summary and Conclusions

- Nerve arterioles dilate and constrict during continuous vibration exposure.
- This phenomenon is similar to the cycling observed in the rat tail artery and tail skin perfusion, although the cycling is out of phase.
- The presumed cyclic reductions and elevations in blood flow to the nerve during vibration may play a role in causing nerve edema and injury.

References

Curry BD, Bain JLW, Yan J-G, Zhang LL, Yamaguchi M, Matloub HS, Riley DA [2002]. Vibration injury damages arterial endothelial cells. Muscle Nerve 25(4):527–534.

Govindaraju SR, Bain J, Riley DA [2007a]. Effects of vibration exposure on blood flow to the rat-tail skin. In: Proceedings of the 11th International Conference on Hand-Arm Vibration (Bologna, Italy, June 3–7, 2007), pp. 239–243.

Govindaraju SR, Bain JLW, Riley DA [2007b]. Vibration causes ischemia-reperfusion injury in the rat-tail artery. FASEB J 21:899.10.

Govindaraju SR, Curry BD, Bain JLW, Riley DA [in press]. Nerve damage occurs at a wide range of vibration frequencies. Int J Ind Ergon.

REDUCED SENSORY NERVE THRESHOLDS MAY BE INDICATIVE OF NERVE INJURY AFTER REPEATED EXPOSURE TO VIBRATION

Kristine Krajnak, G. Roger Miller, Claud Johnson, and Stacey Waugh

Engineering and Control Technology Branch, Health Effects Laboratory Division, National Institute for Occupational Safety and Health, Morgantown, WV

Introduction

Hand-arm vibration syndrome (HAVS) is characterized by the occurrence of cold-induced vasospasms and a reduction in tactile sensitivity. Although vascular dysfunction usually improves after workers stop using vibrating handtools, sensory symptoms are often maintained or degrade over time [Nasu and Ishida 1986]. If changes in nerve function could be detected earlier, it may be possible to intervene or treat workers and eliminate further damage to the sensory system. The goal of this study was to characterize changes in sensory nerve function during a 25-day exposure to vibration in a rat tail model of HAVS. Changes in sensory nerve function were assessed using the current perception threshold (CPT) test.

Methods

Animals. Male Sprague Dawley rats (6 weeks of age) were housed in AAALAC-accredited facilities. All procedures were approved by the NIOSH Animal Care and Use Committee and were in compliance with CDC guidelines for the care and use of laboratory animals. Vibration exposures were performed by restraining rats in a Broome-style restrainer and securing their tails to the vibration platform. Restraint control animals were treated in an identical manner except that the tail platform was set on isolation blocks instead of a shaker. Rats were exposed to daily bouts of vibration (4 hr/day, 125 Hz, constant acceleration 49 m/s^2 root-mean-square) or restraint for 25 consecutive days. An additional group of animals served as cage control rats. These animals were tested at the same times as the other animals, but they were not exposed to restraint or vibration. All animals were euthanized with an overdose of pentobarbital (100 mg/kg), and tail nerves were collected for 2′-3′ cyclic nucleotide 3′ phosphodiesterase (CNPase) immunohistochemical analyses. CNPase levels are positively correlated with glia-axon contact in peripheral nerves [Toma et al. 2007]. CNPase staining was assessed in the right ventral nerve bundle (i.e., large nerve) and in smaller nerve bundles (i.e., small nerves less than 100 μm in diameter) found around the ventral artery. The labeled area of each nerve bundle was measured using densitometry (Scion Image).

CPT tests. Sensory neuron function was assessed by measuring CPTs with a Neurometer (Neurotron, Inc., Baltimore, MD). Transcutaneous nerve stimulation was applied to the C10 region of the tail. Three frequencies were used to test specific fiber types (5 Hz – C; 250 Hz – Aδ; and 2,000 Hz – Aβ). The intensity of the stimulus was automatically increased in small increments until the rat flicked its tail. Tests at each frequency were repeated until the animals displayed two responses that were within 2 CPT (or 0.02 mA) of each other (two to three tests per animal). CPT tests were performed after vibration exposure on the first day of each week.

Data analyses. Sensory nerve thresholds were analyzed using a mixed model three-way ANOVA where the independent variables were treatment, days of exposure, and pre/post exposure. Animal served as a random variable for each analysis. Biological data were analyzed using a one-way ANOVA where treatment was the independent variable. Differences with $p < 0.05$ were considered significant.

Results

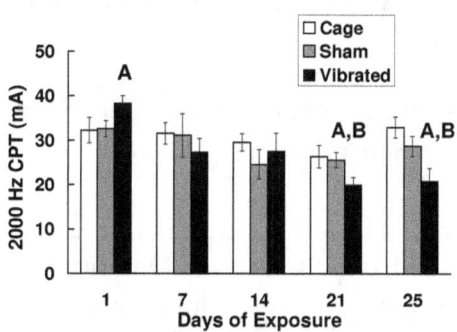 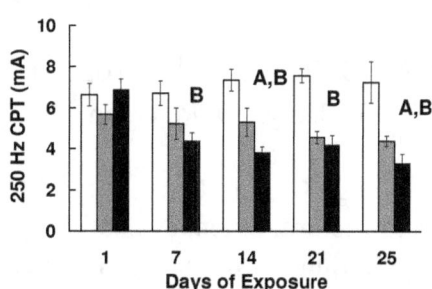

Figure 1.—Weekly CPTs in rat tails exposed to 25 days of vibration. Exposure to vibration resulted in a reduction in 2,000- and 250-Hz thresholds over time (A: different from cage controls; B: different from day 1 vibrated, $p < 0.05$). Restraint controls also displayed lower 250-Hz CPTs than cage controls after 21 and 25 days of exposure.

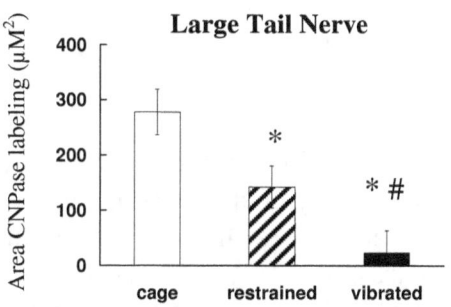 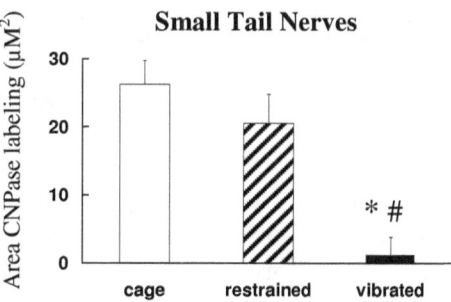

Figure 2.—The area of CNPase immunostaining in large and small tail nerves of rats. CNPase immunolabeling was lower in the large, but not small tail nerve bundles of rats exposed to restraint control conditions (* less than cage control, $p < 0.05$; # less than restraint control, $p < 0.05$).

Discussion

- CPTs in Aβ fibers (2,000 Hz) decline with repeated exposure to vibration. The reduced thresholds of Aβ fibers to stimulation could potentially lead to an increased sensitivity of rats to mechanical stimuli and serve as an early indicator of nerve injury.
- Vibrated and restraint control rats displayed reductions in the 250-Hz CPT, indicating that changes in Aδ fiber sensitivity are related to restraint and not necessarily due to vibration.
- Vibration-induced reductions in the 2,000-Hz thresholds were associated with a reduction in area labeled with CNPase in large and small nerve bundles. A reduction in the density of this enzyme shown is associated with reduced glia-axon contact in peripheral nerves and may indicate nerve dysfunction [Toma et al. 2007].

References

Nasu Y, Ishida K [1986]. Follow-up study of patients with vibration syndrome in Japan. Scand J Work Environ Health *12*:313–319.

Toma JS, McPhail LT, Ramer MS [2007]. Differential RIP antigen (CNPase) expression in peripheral ensheathing glia. Brain Res *1137*:1–10.

INFLUENCE OF CHANGE IN WORK PRACTICES ON VIBROTACTILE PERCEPTION: PRELIMINARY ANALYSIS OF A PROSPECTIVE STUDY OF FORESTRY WORKERS

A. J. Brammer,[1,2] M. G. Cherniack,[1] E. Toppila,[3,4] P. Sutinen,[3,5] R. Lundstrom,[6] T. Nilsson,[7] G. Neely,[8] D. Peterson,[1] N. Warren,[1] and T. Morse[1]

[1]Ergonomic Technology Center, University of Connecticut Health Center, Farmington, CT
[2]Institute for Microstructural Sciences, National Research Council, Ottawa, Ontario, Canada
[3]Department of Otorhinolaryngology, Tampere University Hospital, Tampere, Finland
[4]Finnish Institute of Occupational Health, Helsinki, Finland
[5]Department of Physical Medicine and Rehabilitation, North Karelia Central Hospital, Joensuu, Finland
[6]Department of Biomedical Engineering and Informatics, University Hospital, Umeå, Sweden
[7]Department of Occupational and Environmental Medicine, Sundsvall Hospital, Sundsvall, Sweden
[8]Department of Psychology, Umeå University, Umeå, Sweden

Introduction

The specification of tolerable exposures of the hand to vibration, in particular exposures that precipitate few or no neurosensory symptoms, remains a primary goal for occupational health standards and regulations. Numbness in the fingers or hands, paresthesias, and a reduction in handgrip or difficulty performing fine manipulative tasks are commonly experienced by manual workers who operate power tools and have been linked to vibrotactile perception thresholds (VPTs) [Coutu-Wakulczyk et al. 1997]. The purpose of the present work was to apply a recently developed metric derived from VPTs to explore the consequences of a change in work practices of forestry workers associated with the introduction of mechanical tree harvesting (in 1999). The change reduced the exposure to vibration by substituting brush cutters for chain saws.

Methods

A 13-year prospective study was conducted on a subgroup of an open cohort of forestry workers for evidence of work-related changes in hand and arm function. At baseline the subjects constituted almost 20% of the open cohort: 28% of the cohort, and subgroup, reported neurosensory symptoms in 1990. The full cohort and subgroup were reexamined in 1995 and 2003. The VPTs were determined at the fingertips of digits 3 and 5 of both hands at stimulus frequencies of 4 and 6.3 Hz (believed mediated by the SAI receptors) and 20 and 32 Hz (believed mediated by the FAI receptors) using apparatus complying with ISO 13091-1 (Method A). The metric, which has been shown to be applicable to chain saw operators, involves constructing the mean change in VPT per receptor population at each fingertip over a time interval of months or years, and then summing the threshold changes across the SAI and FAI receptor populations at each fingertip [Brammer et al. 2007a,b]. The metric is insensitive to the age of the subject. The statistical significance of the observed threshold change is derived from the known threshold test-retest repeatability for the apparatus and measurement algorithm. The summed threshold changes were formed for the intervals 1990–1995 and 1995–2003, and trends between the values for the two intervals were then analyzed.

Results

Of the 23 persons (23/124) admitted to the study in 1990, 18 (18/109) returned for reexamination in 1995 and 10 (10/59) in 2003. Only one subject (#1) reported neurosensory symptoms in 2003. Table 1 shows that of the 20 hands examined on all three occasions, 16 were exposed to vibration (subjects #2 and #9 were foremen). Of the 20 hands, 10 were assessed to be improving in acuity, 5 deteriorating in acuity, and 4 unchanged. The VPTs from one hand gave conflicting results (left hand of subject #1). For the vibration-exposed workers, the trends were as follows: eight hands were assessed to be improving in acuity, five deteriorating in acuity, and two unchanged.

Table 1.—Trend of summed threshold changes between 1990–1995 and 1995–2003

Subject #	Age (years), 2003	Exposed to vibration?	Acuity assessment	
			Left hand	Right hand
1	54	Yes	Conflicting results	Deteriorating
2	49	No	Improving	Improving
3	46	Yes	No change	Improving
4	55	Yes	Deteriorating	Deteriorating
5	39	Yes	Improving	No change
6	39	Yes	Deteriorating	Deteriorating
7	48	Yes	Improving	Improving
8	48	Yes	Improving	Improving
9	51	No	No change	No change
10	51	Yes	Improving	Improving

Discussion

The trend in VPTs from the interval before the change in work practices to the interval containing the new work practices is encouraging. While not all subjects seem to be benefiting from the change in work practices, substantially more exhibit improving sensory acuity than deteriorating acuity, and fewer now report neurosensory symptoms. It will be important to follow the forestry workers who experience only the new work practices to establish whether or not their exposure is below a threshold for the onset of neurosensory symptoms.

References

Brammer AJ, Piercy JE, Pyykkö I, Toppila E, Starck J [2007a]. Method for detecting small changes in vibrotactile perception threshold related to tactile acuity. J Acoust Soc Am *121*(2):1238–1247.

Brammer AJ, Sutinen P, Diva UA, Pyykkö I, Toppila E, Starck J [2007b]. Application of metrics constructed from vibrotactile thresholds to the assessment of tactile sensory changes in the hands. J Acoust Soc Am *122*(6):3732–3742.

Coutu-Wakulczyk G, Brammer AJ, Piercy JE [1997]. Association between a quantitative measure of tactile acuity and hand symptoms reported by operators of power tools. J Hand Surg [Am] *22A*:873–881.

This work was supported by the National Institute for Occupational Safety and Health, grant U01 OH 071312.

INDIVIDUAL VARIABILITY IN CHANGES IN FINGER BLOOD FLOW INDUCED BY HAND-TRANSMITTED VIBRATION

Ying Ye and Michael J. Griffin
Human Factors Research Unit, Institute of Sound and Vibration Research
University of Southampton, Southampton, U.K.

Introduction

Finger blood flow (FBF) is influenced by many factors, including body temperature and hand-transmitted vibration. This study investigated the manner in which the reduction in FBF caused by vibration depends on room and finger temperature in individual subjects.

Methods

FBF was measured in 12 healthy male subjects in two different room temperatures (20 ± 1 °C and 28 ± 1 °C) on two separate days, with the order of conditions balanced over subjects. FBF was measured using strain gauge venous occlusion plethysmography according to the technique proposed by Greenfield et al. [1963]. A plastic cuff was fitted around the proximal phalanx of the left middle finger, with the soft plastic tube from the cuff connected to an *HVLab* Multi-Channel Plethysmograph (University of Southampton, U.K.). A mercury-in-silastic strain gauge was placed around the finger at the baseline of the nail. Finger skin temperature (FST) was measured by a k-type thermocouple attached by micropore tape to the distal phalanx of the left middle finger. Room temperature was also measured using a k-type thermocouple.

Subjects lay supine throughout the experiment with both hands supported at heart level. After 15 min acclimatization, FBF and FST were measured on the middle finger of the left hand every minute while the right hand applied a downward force of 5 N. During the second of two 10-min periods, the right hand was exposed to sinusoidal 125-Hz vertical vibration at 44 ms^{-2} rms (unweighted).

Nonparametric tests (Wilcoxon matched-pairs signed-ranks test for two related samples and the Spearman rank-order correlation) were used in the statistical analysis.

Results

The higher room temperature (28 °C) resulted in greater FST ($p = 0.001$) and greater FBF ($p = 0.002$) than the lower room temperature (20 °C). At both temperatures, FBF reduced during vibration ($p < 0.001$). The absolute reduction in FBF differed between the two room temperatures ($p < 0.01$), but the percentage reduction relative to FBF before vibration was not significantly different ($p = 0.436$).

For each subject, Figure 1 shows the median FBF during the 10-min periods before and during vibration with the room temperature at 20 °C and 28 °C. Subjects with a high FBF before vibration show greater absolute reduction in FBF during vibration.

The relation between the median individual FBFs before and during vibration at both room temperatures is shown in Figure 2. The percentage reduction in FBF caused by vibration is similar in all cases, even though subjects have different FBFs due to individual factors and different room temperatures.

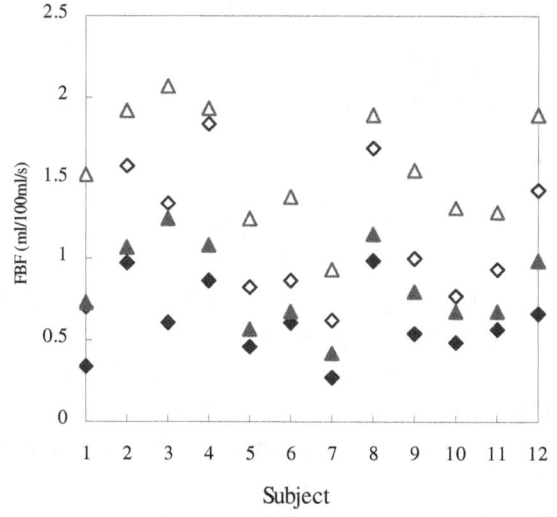

Figure 1.—Individual median FBF during 10 min in unexposed left finger before and during vibration of right hand. Open diamond = before vibration (20 °C); solid diamond = during vibration (20 °C); open triangle = before vibration (28 °C); solid triangle = during vibration (28 °C).

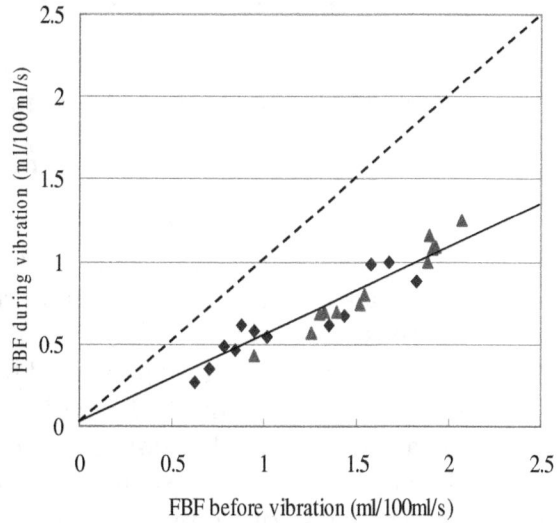

Figure 2.—Individual median FBF before and during vibration. Diamond = 20 °C; triangle = 28 °C.

During vibration, the variability in FBF (standard deviation measured over 10 min) was correlated with the mean FBF ($p < 0.001$ at 20 °C; $p < 0.001$ at 28 °C). The coefficient of variation in FBF within individuals during vibration at 28 °C was significantly greater than at 20 °C ($p = 0.014$), but with no correlation in the coefficient of variation between temperatures ($p = 0.796$).

The FST was correlated with FBF before vibration ($p < 0.001$ at 20 °C; $p < 0.001$ at 28 °C), but not during vibration ($p = 0.093$ at 20 °C; $p = 0.067$ at 28 °C). At 20 °C and 28 °C, FBF and FST before vibration were correlated with FBF and FST, respectively, during vibration ($p < 0.005$).

Before and during vibration, both FBF and FST were correlated between 20 °C and 28 °C (FBF-before: $p < 0.001$; FST-before: $p = 0.006$; FBF-during: $p = 0.046$; FST-during: $p = 0.037$).

Conclusions

Individual FBF is influenced by many factors, including room temperature, finger temperature, and hand-transmitted vibration. Increased room temperature and increased finger temperature increase FBF. Hand-transmitted vibration seems to produce a reduction in FBF that is proportional to the absolute FBF.

Reference

Greenfield ADM, Whitney RJ, Mowbray JF [1963]. Methods for the investigation of peripheral blood flow. Br Med Bull *19*:101–109.

VIBRATION EXPOSURE IN OFFICES: A CASE STUDY

Stana Živanović and Aleksandar Pavić
Department of Civil and Structural Engineering, University of Sheffield, Sheffield, U.K.

Introduction

Buildings are constantly exposed to various kinds of vibration during their lifetime (such as traffic, ground-borne vibration, wind, and human-induced vibration). Typical receivers of these vibrations are humans occupying the structure. Apart from guidelines devoted specifically to human response to vibration [BSI 1992], there are vibration serviceability design guidelines for floor structures (such as those published by the U.K. Concrete Society [Pavić and Willford 2005]) that define vibration limits depending on the intended use of the structure (hospital operating theaters, residential buildings, offices, or workshops). Civil engineers typically prefer to use guidelines linked to the design of buildings rather than those devoted exclusively to vibration perception because of simplicity of the former and unfamiliarity with the latter.

Data related to in-service vibration of a building are valuable, but are rare. This abstract presents such data from 1 day of monitoring of vertical vibration (generated by human walking) of an office building floor. The collected data were processed to calculate the vibration measures: vibration dose value (VDV), root-mean-square (rms), and R factor. These are discussed in the context that the office floor was subject to complaints due to its excessive vibration during normal occupancy.

Description of Building and Monitoring Process

The building investigated contains offices. On one of its suspended floors, adverse comments about vibration had been reported by the occupants, especially at a particular work station. The vertical acceleration level generated by people walking on the floor during their normal work activities was measured by a piezoelectric accelerometer placed close to the work station. The accelerometer and the acquisition PC (controlled remotely) were hidden in the false floor; therefore, the office staff was not aware of their presence or position.

Results

The acquired acceleration was measured during 12 hr and weighted using the Wb weighting curve [BSI 1992]. The eight working hours (9.00 a.m.-5:00 p.m.) of the weighted record were extracted (Figure 1a). 1s rms trend is presented in the same figure. 1s rms trend was divided by the base acceleration level of 0.005 m/s^2 to obtain response factors (or R factors). R factors are typically used for vibration serviceability assessment of buildings; for office buildings, the maximum R factor should not exceed a value of 4 [Pavić and Willford 2005]. Figure 1b shows that, for the data analyzed, R factor exceeds this value very rarely (only 0.7% of the total time). Since people complained about the in-service vibration, this limiting value might not be appropriate.

In contrast with the serviceability guideline, BS 6472 [BSI 1992] requires evaluation of vibration using the VDV if the crest factor is above 6. To calculate the crest factor and to study the vibration dependence on exposure time, the rms acceleration for different time windows was calculated first (Figure 1c). The value stabilized around 0.0066 m/s^2 only when the observation time was longer than 100 min. By dividing the corresponding peak acceleration by the rms value, the dependence of the crest factor on observation time was obtained as well (Figure 1d). This value varied depending on observation time, suggesting that choice of evaluation time should be made carefully. However, in this case, the crest factor was always greater than 6 except for very small time windows. Therefore, the VDV should be used to evaluate the human response [BSI 1992]. During eight working hours, this value reached 0.146 m/s$^{1.75}$. However, no VDV limit for an office building is defined in the current guidelines.

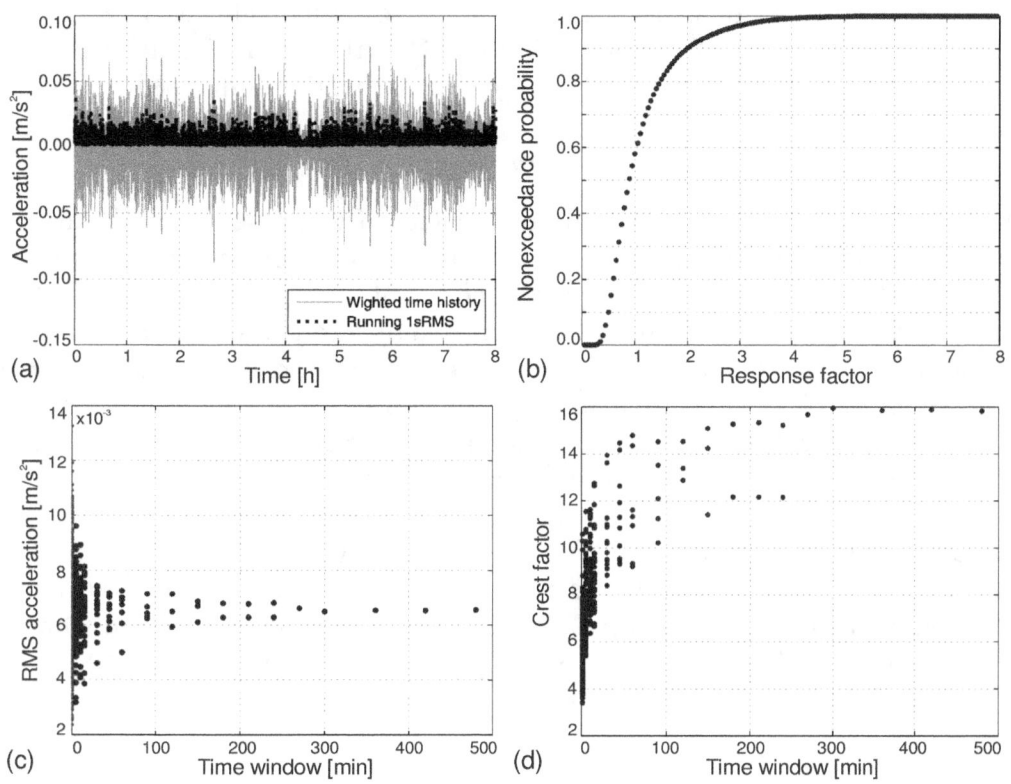

Figure 1.—(a) 8-hr weighted time history; (b) cumulative probability of response factor; (c) rms acceleration; (d) crest factor as a function of time window.

Discussion

Based on in-service vibration monitoring in an office building (subject to complaints about its vibration level), it was found that the current vibration limit of R=4 is probably set too high. BS 6472 suggests that the VDV should be used for evaluating measured type of vibration rather than the R factor, but it does not define the VDV limit for an office building. This suggests that additional research in the field is required to help civil and structural engineers design buildings against excessive vibration induced by human walking.

References

BSI [1992]. Guide to evaluation of human exposure to vibration in buildings (1 Hz to 80 Hz). London: BSI British Standards. BS 6472:1992.

Pavić A, Willford M [2005]. Vibration serviceability of post-tensioned concrete floors. In: Post-tensioned concrete floors – design handbook. 2nd ed. Technical report 43. Surrey, U.K.: The Concrete Society, pp. 99–107.

DETERMINATION OF OSTEOGENIC FREQUENCIES RELATED TO THE OSSEOINTEGRATION OF IMPLANTED DEVICES

Joseph Starkman,[1] Farid Amirouche,[2] and Mark Gonzalez[3]

[1]Midwestern University, Chicago College of Osteopathic Medicine
[2]University of Illinois at Chicago, Department of Bioengineering
[3]University of Illinois at Chicago, Department of Orthopaedic Surgery

Summary

The goal of this research is to identify frequencies of vibration that promote the development of bone in-growth as it relates to the stability of implanted devices such as those found in a total hip arthroplasty. There are both therapeutic and harmful levels of vibrations that affect the human body. Noxious levels of vibration are considered by frequency measured in hertz, magnitude measured in g-force, and duration of exposure. Pathological responses are known to occur at high frequencies (5–100 Hz) and large magnitudes (>1 g). When vibration is experienced, relatively low-peak g levels can be severely damaging if they are at the resonant frequency of organs and connective tissues [BME at Stony Brook 2006]. There are no data to suggest noxious effects in the 15–50 Hz range at 0.56 g. The ISO standard is 0.8 g at 20–50 Hz for no more than 0.5 hr. Recent whole-body vibration therapy (WBVT) devices involve subjecting stationary individuals to vibrating platforms at certain frequencies and amplitudes of forced vibrations (Figure 1). One recent study used frequencies of 35–40 Hz and 2.28–5.09 g. The results showed vibration training improved isometric and dynamic muscle strength and increased bone mineral density (BMD) of the hip compared to control groups [Verschueren et al. 2004].

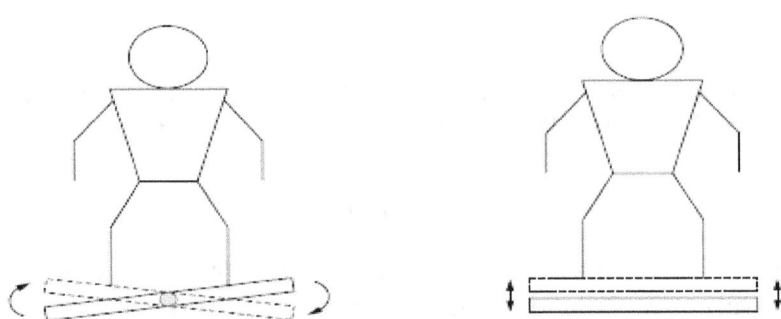

Figure 1.—Oscillating versus vertical vibrating plates.

Suboptimal long-term stability of total hip arthroplastic devices has been attributed to poor bone in-growth into acetabular components (i.e., osseointegration) mainly due to insufficient subchondral vascular supply [Morscher et al. 2002]. Various improvements in materials and surgical approaches have led to increased stability of acetabular components over the years, but long-term stability remains an issue. Rapid fluid flow along with the effects of piezoelectricity and streaming potentials affect bone cellular processes in the relative absence of significant mechanical loading. For example, tiny acceleratory motions of 0.3 g at 45 Hz for 10 min/day have been shown to sufficiently influence bone formation and bone morphology [Garman et al. 2007]. Given the recent data that tiny acceleratory motions independent of direct loading of the matrix have been shown to be anabolic to skeletal tissue and that vascular-induced interstitial fluid flow (IFF) is known to be continuously present regardless of mechanical loading [Hillsley et al. 1993], it is theorized that nonmechanical loading mechanisms and vascular IFF play a larger role in bone remodeling than previously thought. Non-load-bearing experiments involving the use of pulsed electromagnetic field (PEMF) frequencies close to 15 Hz (1 G (0.1 µT); electric field strength 2 mV/cm) point toward the involvement of electromagnetically induced mechanisms that significantly contribute to osteogenesis [Massari et al. 2006; Chang et al. 2004]. It is hypothesized that combined therapy using vibration devices and PEMFs stimulate osteogenesis through enhancement of piezoelectric, streaming potential and other fluid regulatory forces within collagenous structures. The alleviation of soft tissue restrictions improves muscle tone, increases bioelectric signals between cells, and promotes greater IFF among cells. Improvement of the communication between cells via restoration of bioelectric conduction improves the inherent cellular processes and leads to greater bone stability. It may also lead to better in-growth of implanted materials.

The exact frequencies and amplitudes that stimulate therapeutic levels of vibration (e.g., improved BMD and bone in-growth of acetabular components) without causing long-term cellular damage or damage to implanted devices are not known and require further investigation. This research aims to discover the osteogenic frequencies and develop a method of controlled application of such vibrations. Eventually, therapeutic protocols implementing various osteogenic frequencies (i.e., both PEMF frequencies and WBVT) to rehabilitation therapies for acute and chronic bone-involving conditions (e.g., fractures and osteoporosis) may be developed from this research.

References

BME at Stony Brook [2006]. Contraindications and potential dangers of the use of vibration as a treatment for osteoporosis and other musculoskeletal diseases. Stony Brook, NY: Department of Biomedical Engineering, Stony Brook University, January 11, 2006. [http://www.bme.sunysb.edu/bme/people/faculty/docs/crubin/safety-1-11-06.pdf]. Date accessed: May 2008.

Chang WH, Chen LT, Sun JS, Lin FH [2004]. Effect of pulse-burst electromagnetic field stimulation on osteoblast cell activities. Bioelectromagnetics *25*(6):457–465.

Garman R, Gaudette G, Donahue LR, Rubin C, Judex S [2007]. Low-level accelerations applied in the absence of weight bearing can enhance trabecular bone formation. J Orthop Res *25*(6):732–740.

Hillsley MV, Frangos JA [1993]. Bone tissue engineering: the role of interstitial fluid flow. Biotechnol Bioeng *43*(7):573–581.

Massari L, Fini M, Cadossi R, Setti S, Traina GC [2006]. Biophysical stimulation with pulsed electromagnetic fields in osteonecrosis of the femoral head. J Bone Joint Surg Am *88*(Suppl 3):56–60.

Morscher EW, Widmer KH, Bereiter H, Elke R, Schenk R [2002]. Cementless socket fixation based on the "press-fit" concept in total hip joint arthroplasty (in German). Acta Chir Orthop Traumatol Cech *69*(1):8–15.

Verschueren SM, Roelants M, Delecluse C, Swinnen S, Vanderschueren D, Boonen S [2004]. Effect of 6-month whole body vibration training on hip density, muscle strength, and postural control in postmenopausal women: a randomized controlled pilot study. J Bone Miner Res *19*(3):352–359.

VIBRATION GUIDELINES: OVERREGULATION IN EUROPE PLACES AN UNJUSTIFIED BURDEN ON THE CONSTRUCTION INDUSTRY

Michael Fischer[1] and Todd Lutz[2]

[1]Wacker Neuson SE, Munich, Germany
[2]Wacker Corp., Menomonee Falls, WI

Introduction

With a cost of several hundred million euros per year, European Directive 2002/44/EC places an unjustified burden on the European construction industry. Equipment owners are forced to rate the vibration dose for every employee. In Germany, approximately 90,000 small construction businesses are affected, many with fewer than 20 workers. A basic issue with European legislation is that through national transformation laws, regulations can become more restrictive. For example, in Germany the whole-body vibration (WBV) limit for the z direction was reduced from 1.15 to 0.8 m/s^2. Industry's concerns that there is no justification behind such a change were not taken into account, and the basic goal of European harmonization is in danger. During the last decade, the number of occupational diseases has continually decreased through industry efforts. Thus, it is the position of Wacker Neuson that the vibration guidelines for both hand-arm vibration (HAV) and WBV are not necessary and should be modified or abolished.

Methods

With respect to worker health and safety, Wacker Neuson gathered statistical data from the Berufsgenossenschaften (BG) (institutions for statutory accident insurance and prevention in Germany). BG provided detailed data on confirmed cases of occupational diseases for both HAV and WBV. With regard to measurements, we are able to record data on our own test stands. We have extensive experience in measuring vibratory plates, rammers, breakers, wheel loaders, and other tools that are provided for the construction industry. We also took into account literature from highly regarded experts in the field of vibration assessment.

Results

For Germany's construction industry, we asked for official data regarding the number of occupational diseases caused by HAV and WBV. For white-finger syndrome, with the help of HVBG (federation of commercial professional associations) [Griesmeier-Goller 2005], we found only two confirmed cases in 2004, while the trend from prior years showed a constant decrease. The same result was found for occupational diseases caused by WBV, where only five cases were confirmed in 2004 with a similar trend for prior years. The numbers clearly demonstrate that industry's efforts to reduce HAV and WBV, driven by market requirements, have yielded excellent results. No additional regulation is needed to require the end user to follow highly bureaucratic evaluation procedures.

We also question the experimental methods that are used to obtain vibration results. For example, with regard to HAV on vibratory plates, we find that the standard measurement procedure given in EN 500-4 shows a scatter of ±1 to 2 m/s^2 with increasing size of the compaction device. Schenk and Gillmeister [2004] showed through tests of HAV values for 12 handheld construction tools by 18 separate laboratories that data variations occurred for 40%–50% of these institutions. In this study, the measurements of the only ISO 9001-certified laboratory were disregarded because the lab's values deviated too far from the average. If the requirement is below 2.5 m/s^2 and 5 m/s^2 to overcome the need of daily reporting requirements, then the issue of data scatter is critical. A machine measured with, say, 4 m/s^2 by a market surveillance authority (which currently does not exist) could not be rated precisely. The supplier could properly claim the HAV to be above 5 m/s^2 or below 2.5 m/s^2, and nobody can prove the opposite.

In discussions with health experts (see also Tominaga [2005]), we found that the frequency weighting curve used in HAV measurements is outdated and, as presented at the Human Vibration Conference held in Las Vegas, NV, in 2004, its accuracy relative to health issues is questionable. For instance, darkness, cold, smoking, and many other conditions can affect white-finger syndrome, even in the absence of vibration. Subsequent medical studies have demonstrated that the weighting factor used in the standard is weighted toward the perception of vibration, not to the frequencies causing the greatest neurovascular damage. Coupling of the fingers to the handgrip is also an issue for both measurement and health assessment [BG 2004]. When one rides a bicycle downhill, one does not grip the handlebars as tightly as possible. Instead, one grips at a minimal level that feels comfortable and the bike remains controllable. The same is true for vibratory plate users. Most often, the operator follows the self-driven machine instead of maintaining a tight grip on the handle.

For wheel loaders, there is no procedure to determine WBV. Statistical determinations of WBV on construction sites are listed in ISO Technical Report (TR) 25398, with discussion of raising it from the level of a technical report to a standard. By doing this, ISO regards the state of the art to be sufficient regarding health and safety of the operator. Obviously no regulation is needed, as industry has already demonstrated success with its market-driven efforts in a competitive environment.

Discussion

Wacker Neuson takes the position that the European vibration regulation and related documents are not necessary and should be revised or abolished. The financial burden on the European construction industry is several hundred million euros per year. This expenditure is not justified by the limited and even declining number of reported HAV or WBV injuries. Methods to measure levels of HAV and WBV are questionable at best, and an industry surveillance authority does not exist. Industry, in response to the market, has been the driving force for technical progress regarding the health and safety of the end user. It is recommended that a similar arbitrary and inaccurate standard *not* be established in other countries as was done in the European Community.

References

BG (Die gewerblichen Berufsgenossenschaften) [2004]. Report on project No. BGIA 4098 – Hand-arm vibrations and coupling forces: measurement methods (in German).

Griesmeier-Goller [2005]. TBG, HVBG, personal communication. November 9.

Schenk T, Gillmeister F [2004]. Measurement uncertainties of vibration-emitting hand-held tools (in German). Düsseldorf, Germany: VDI report No. 1821, pp. 97–114.

Tominaga Y [2005]. New frequency weighting of hand-arm vibration. Ind Health (Japan) *43*(3):509–515.

Session X: Whole-Body Vibration III

Chair: Alan Mayton
Co-Chair: Miyuki Morioka

Presenter	Title	Page
M. Morioka University of Southampton	Dependence of Equivalent Comfort Contours for Vertical Vibration of the Foot on Vibration Magnitude	111
M. L. M. Duarte * Federal University of Minas Gerais	Combined Effects of Whole-Body Vibration Exposure and Noise on Human Temporary Threshold Shift	113
P. W. Johnson University of Washington	A Comparison of Whole-Body Vibration Exposures Between Low- and High-Floor Buses	115
T. Eger Laurentian University	Changes in Seat-Head Whole-Body Vibration Transmissibility and Muscle Activity Under Assymetric Neck and Trunk Postures	117
J. P. Dickey University of Guelph	The Nature of Multiaxis Six-Degree-of-Freedom Vehicle Vibrations in Forestry, Mining, and Construction Heavy Equipment	119

* Author unable to attend conference.

DEPENDENCE OF EQUIVALENT COMFORT CONTOURS FOR VERTICAL VIBRATION OF THE FOOT ON VIBRATION MAGNITUDE

Miyuki Morioka and Michael J. Griffin
Human Factors Research Unit, Institute of Sound and Vibration Research
University of Southampton, Southampton, U.K.

Introduction

In transport, at workplaces, and during leisure and domestic activities, vibration is transmitted to the body via the feet, as well as the seat and hands, and can cause discomfort and annoyance. To evaluate vibration at the feet, ISO standard 2631-1 (1997) uses frequency weighting W_k, while British Standard 6841 (1987) uses frequency weighting W_b. Frequency weighting W_b was influenced by equivalent comfort contours for vibration of the feet at vibration magnitudes equivalent to 10-Hz vertical seat vibration at 0.8 ms^{-2} root-mean-square (rms) [Parsons et al. 1982]. Recent studies suggest that the equivalent comfort contours may depend on vibration magnitude, as with whole-body vibration [Morioka and Griffin 2006].

This experiment was designed to determine the effect of vibration magnitude, from the threshold of perception to levels associated with severe discomfort, on equivalent comfort contours for vertical vibration at the foot over the frequency range 8–315 Hz.

Methods

Twelve males (mean age = 24.9 years, mean stature = 177.1 cm, mean weight = 73.1 kg) were exposed to vertical vibration at the foot via a wooden footrest (30.5 by 10.5 mm with 10° inclination) mounted rigidly to a vibrator (Derritron VP30).

Subjective magnitudes for the vibration of the foot were determined using the method of magnitude estimation. Subjects compared pairs of motions (a reference and a test motion), each lasting 2 sec and separated by 1 sec. The reference motion was fixed at 5.0 ms^{-2} rms at 50 Hz. The test motions were randomly selected from a range of frequencies (8–315 Hz in one-third octave steps) and a range of magnitudes (0.002–0.126 ms^{-1} rms velocity in 3-dB steps, but limited to 100 ms^{-2} rms). The task was to assign a number that represented the discomfort of the test motion, assuming the discomfort of the reference motion was 100. Absolute perception thresholds were also determined for frequencies from 8 to 315 Hz and are reported elsewhere [Morioka and Griffin, forthcoming].

The relationship between the sensation magnitude, ψ, and the vibration magnitude, φ, was determined for each frequency using Stevens' power law with an additive constant assuming no sensation below the perception threshold:

$$\psi = k(\varphi - \varphi_0)^n \qquad (1)$$

where k is a constant, φ_0 is the perception threshold, and n is the rate of change of the logarithm of sensation with the logarithm of vibration magnitude. Linear regression was performed using:

$$\log_{10}\psi = n\log_{10}(\varphi - \varphi_0) + \log_{10}k \qquad (2)$$

Results

The calculated vibration magnitude, φ (in acceleration), is shown as a function of vibration frequency for comfort contours equivalent to sensation magnitudes from 25 to 300 in Figure 1. It can be seen that the shapes of the equivalent comfort contours depend on the sensation magnitude, reflecting a nonlinear response. With high-sensation magnitudes, the comfort contours are approximately contours of constant velocity. With low-sensation magnitudes, the contours become similar to the absolute perception threshold.

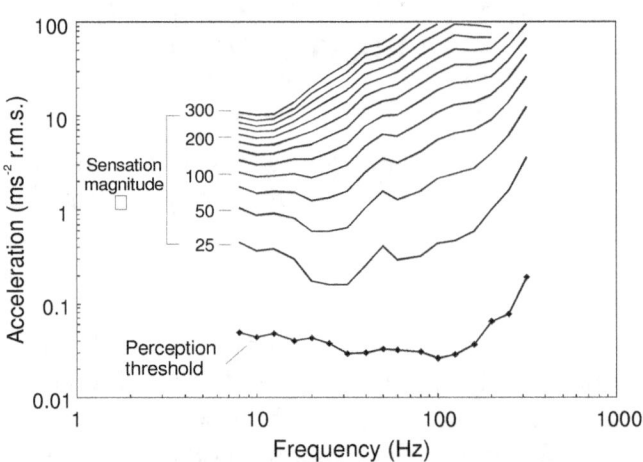

Figure 1.—Equivalent comfort contours for vertical foot vibration at sensation magnitudes from 25 to 300 (100 is equivalent to 5.0 ms^{-2} rms at 50 Hz).

Discussion

The equivalent comfort contours for vertical vibration of the foot were inverted and normalized to a value of unity at 8 Hz and then overlaid with the W_b and W_k frequency weightings, as shown in Figure 2. The standardized frequency weightings underestimate the discomfort of low-magnitude vibration at frequencies greater than about 30 Hz (or, conversely, overestimate the discomfort caused by frequencies less than 30 Hz). The magnitude-dependence of the equivalent comfort contours means that no single linear frequency weighting provides an accurate prediction of discomfort caused by vibration at the feet over the range of vibration frequencies and magnitudes investigated.

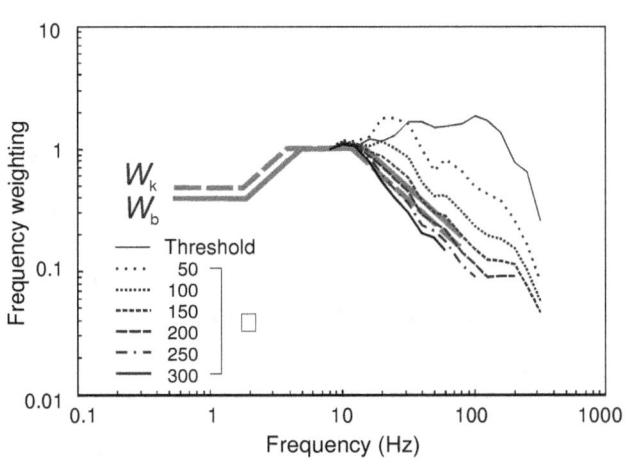

Figure 2.—Effect of vibration magnitude on frequency weightings (inverted equivalent comfort contours normalized at 8 Hz).

References

BS 6841 [1987]. Guide to measurement and evaluation of human exposure to whole-body mechanical vibration and repeated shock. London: BSI British Standards.

ISO [1997]. Mechanical vibration and shock: evaluation of human exposure to whole-body vibration. Part 1: General requirements. Geneva, Switzerland: International Organization for Standardization. ISO 2631-1:1997.

Morioka M, Griffin MJ [forthcoming]. Absolute thresholds for the perception of fore-and-aft, lateral and vertical vibration at the hand, the seat and the foot. J Sound Vibration.

Morioka M, Griffin MJ [2006]. Magnitude-dependence of equivalent comfort contours for fore-and-aft, lateral and vertical whole-body vibration. J Sound Vibration *298*:755–772.

Parsons KC, Griffin MJ, Whitham EM [1982]. Vibration and comfort. III. Translational vibration of the feet and back. Ergonomics *25*(8):705–719.

COMBINED EFFECTS OF WHOLE-BODY VIBRATION EXPOSURE AND NOISE ON HUMAN TEMPORARY THRESHOLD SHIFT

M. L. M. Duarte, R. Izumi, M. A. R. Santos, E. B. Medeiros, B. C. Siqueira, and L. A. P. de Carvalho

Federal University of Minas Gerais, Mechanical Engineering Department, Group of Acoustics and Vibration in Human Beings, Belo Horizonte/MG, Brazil

Introduction

The most common combination of physical risks in work environments is noise and vibration. The effects of this combination on the health of people have been very little researched and understood, especially the effects on hearing.

Methods

In this laboratory study, our goal was to check the influence of whole-body vibration (WBV) and noise on the temporary threshold shift (TTS) of healthy subjects with respect to the limits set by European Directive 2002/44/EC [2002] and ISO standard 2631-1 (1997). A total of 13 young adult subjects (10 males and 3 females) without any hearing problem and with no history of occupational exposure to noise and/or vibration were used as volunteers (average age of 23.9 ± 5.6 years, average weight of 70.2 ± 10.8 kg, and average height of 1.75 ± 0.07 m). They were submitted to three test conditions: (1) exposure to high sound pressure levels (SPLs) (white noise at 100 dBNA for 15 min), (2) exposure to WBV (z-axis, 6 Hz at 2.45 m/s for 18 min), and (3) combined exposure to high SPL and WBV under the same conditions as before, but with the first starting 3 min after exposure to WBV. The subjects' hearing was evaluated before each test condition using audiometric and distortion product otoacoustic emission (DPOAE) tests [Izumi 2006; Duarte et al. 2007].

Results

The results of the audiometric and DPOAE exams were used as references for the comparison with the exams performed just after each exposure condition described above. It is important to note that the otoacoustic emission results were taken approximately 2 min after the exposures. The comparison of the before and after exposure results aimed to investigate the occurrence of hearing TTS. The results were analyzed statistically using the nonparametric Wilcoxon signed-rank test [Siegel 1975].

This statistical test showed that both the audiometric and DPOAE results after the exposure to noise were worse than those obtained before the exposure, mainly around the 3- and 6-kHz frequencies [Izumi 2006]. Such findings are in accordance with previous studies [Manninen 1983a,b; Nordman et al. 2000; Strasser et al. 2003]. However, for WBV exposure alone, there were no auditory changes in the results, opposing previous studies using different methodologies [Manninen 1983a,b]. The combined exposure showed a big difference in the before and after tests, proving noxious effect of the exposure on human hearing, but no greater than the exposure to noise alone [Izumi 2006].

Discussion

From the results obtained, for WBV alone, there was no significant hearing variation. For the tests using high SPL alone, TTS was present at the 0.5- and 8-kHz frequencies, being the highest variations for both exams at the 3- and 6-kHz frequencies. For the tests using a combination of high SPL and WBV, similar results to noise alone were observed. Quantitatively, only at 4 kHz was the TTS small.

Therefore, for the levels used, the results showed that, in accordance with previous studies, noise has significant effects on the TTS of health subjects. However, no effect was found due to vibration alone. Moreover, the combination of WBV and high SPL had neither additive nor synergic characteristics on TTS. On the contrary, what was observed was a reduction in the hearing effects when compared with the exposure effects of noise alone.

Acknowledgment: This work was developed using funds from Fundação de Amparo à Pesquisa do Estado de Minas Gerais (FAPEMIG) under grant TEC932–05.

References

Directive 2002/44/EC [2002]. Directive 2002/44/EC of the European Parliament and of the Council of 25 June 2002 on the minimum health and safety requirements regarding the exposure of workers to the risks arising from physical agents (vibration).

Duarte MLM, Izumi R, Medeiros EB, de Carvalho LAP, Siqueira BC, Silva WGF, Santos MAR [2007]. The influence of whole-body vibration on the temporary threshold shift of healthy volunteers. In: Proceedings of the 19th International Congress on Mechanical Engineering (Brasilia, Brazil). Paper 1748. CD–ROM.

ISO [1997]. Mechanical vibration and shock: evaluation of human exposure to whole-body vibration. Part 1: General requirements. Geneva, Switzerland: International Organization for Standardization. ISO 2631-1:1997.

Izumi R [2006]. Effects of vibration and high sound pressure levels on temporary threshold shift (Thesis, in Portuguese). Belo Horizonte/MG, Brazil: Federal University of Minas Gerais, Mechanical Engineering Department.

Manninen O [1983a]. Simultaneous effects of sinusoidal whole body vibration and broadband noise on TTS2's and R-wave amplitudes in men at two different dry bulb temperatures. Int Arch Occup Environ Health *51*(4):289–297.

Manninen O [1983b]. Studies of combined effects of sinusoidal whole body vibrations and noise of varying bandwidths and intensities on TTS2 in men. Int Arch Occup Environ Health *51*(3):273–288.

Nordmann AS, Bohne BA, Harding GW [2000]. Histopathological differences between temporary and permanent threshold shift. Hear Res *139*(1):13–30.

Siegel S [1975]. Estatística não-paramétrica para as ciências do comportamento (Nonparametric statistics for the behavioral sciences). São Paolo, Brazil: McGraw-Hill do Brasil Ltda.

Strasser H, Irle H, Legler R [2003]. Temporary hearing threshold shifts and restitution after equivalent exposures to industrial noise and classical music. Noise Health *5*(20):75–84.

A COMPARISON OF WHOLE-BODY VIBRATION EXPOSURES BETWEEN LOW- AND HIGH-FLOOR BUSES

Peter W. Johnson, Jim Ploger, and Ryan Blood
University of Washington, Department of Environmental and Occupational Health Sciences,
Seattle, WA

Introduction

Bus drivers represent a substantial segment of the U.S. transportation workforce. Research has shown an association between exposure to whole-body vibration (WBV) and low back disorders. The goal of this study was to compare and determine whether there were differences in vibration exposures between a low- and high-floor bus. High-floor buses require the rider to ascend steps (typically three) to reach the seating level. In low-floor buses, the rider steps directly on the bus and does not have to climb any steps. For accessibility and efficiency in passenger loading and unloading, low-floor buses are becoming the standard in larger metropolitan areas. Older high-floor buses predominate in many fleets, and in order to fully amortize the cost of the buses, cities are required to keep these buses in service for at least 15 years. As a result, both bus types will be in use as metropolitan companies transition from a predominantly high- to a low-floor bus fleet.

Methods

The study comprised 25 participants: 15 drove a high-floor bus and 10 drove a low-floor bus. The high-floor bus (Gillig Corp., Hayward, CA), typical of somewhat older coach buses, was a 7-year-old, 12.2-m bus with a USSC ALX3 seat. The low-floor bus (New Flyer Industries, Inc., Winnipeg, Manitoba, Canada), typical of newer coach buses in a bus fleet, was a 4-year-old, 12.2-m bus equipped with a new USSC Q91 seat. All participants drove their respective bus for approximately 1 hr over a 65-km standardized test route. The buses were driven under identical load conditions with just the driver (no passengers) on board. The standardized test route included surface streets, freeways, and a small section of road containing eight 4-m speed humps. This route was chosen to represent different types of driving, including start-and-stop driving associated with surface streets, impulsive speed hump excursions, and continuous freeway travel.

The instrumentation developed for this study included a PDA-based portable WBV data acquisition system, which collected raw unweighted WBV data at 640 Hz, and the associated MATLAB and LabVIEW software to analyze WBV exposures, per ISO 2631-1 and 2631-5. Using the serial port on the PDA, global positioning system (GPS) data were also collected and integrated with the WBV exposure data to facilitate the identification of the location (road type) and speed of the bus. The preliminary analysis was focused on analyzing the Z-axis measurements of A_w, vibration dose value (VDV), crest factor, maximum continuous peak, and positive and negative raw weighted peaks. The crest factor was calculated by taking the maximum instantaneous weight peak value encountered during the route and dividing by the A_w for the route. All parameter calculations were identical for the high- and low-floor buses. Data are presented as mean and standard error with significance accepted for p-values less than 0.05.

Results

The GPS data indicated that there were no significant differences in bus speeds between the two bus conditions. Table 1 shows the preliminary results of the analysis of time-weighted average and peak data between the two bus types and between the bus seat and floor. When comparing the floor vibration levels between the buses, the magnitude and direction of the difference depended on the type of measure. Except for the average root-mean-square (rms) vibration exposure (A_w), all floor vibration measures were greater on the low-floor bus, with the impulsive measures (crest factor, maximum peak, and raw weighted peaks) twofold to fourfold higher. When comparing the seat vibration levels, except for the average rms vibration exposure (A_w) and crest factor, the opposite trend in seat vibration levels was observed. All seat vibration measures were lower on the low-floor bus despite the greater floor vibration levels.

Table 1.—Mean and standard error Z-axis vibration measures by location and bus type (n = 10)

	Seat			Floor		
	Low-floor (n = 10)	High-floor (n = 15)	p-value	Low-floor (n = 10)	High-floor (n = 15)	p-value
A_w (m/s^2)	0.40 ± 0.02	0.49 ± 0.01	< 0.001	0.43 ± 0.01	0.46 ± 0.01	0.001
VDV ($m/s^{1.75}$)	9.54 ± 0.47	11.80 ± 0.30	< 0.001	12.37 ± 0.40	10.54 ± 0.23	< 0.001
Crest factor	12.45 ± 0.50	12.38 ± 0.67	0.93	23.73 ± 2.00	12.37 ± 0.53	< 0.001
Maximum peak (m/s^2)	4.71 ± 0.24	5.95 ± 0.32	0.005	9.76 ± 0.76	5.53 ± 0.20	< 0.001
Raw (+) peak (m/s^2)	8.48 ± 1.10	9.65 ± 0.47	0.28	55.31 ± 2.68	15.17 ± 0.60	< 0.001
Raw (−) peak (m/s^2)	−7.14 ± 0.54	−8.89 ± 0.67	0.07	−64.67 ± 5.03	−16.00 ± 1.52	< 0.001

Discussion

There were significant differences in the floor-measured vibration between the buses, with twofold to fourfold higher impulsive vibration measures in the low-floor bus. The opposite trend was seen in the seat; lower vibration levels were measured from the seat in the low-floor bus. These differences could be due to differences in seat design and age. The high-floor bus had a 7-year-old seat, while the low-floor bus a new seat. However, what is evident from this study is that seats on low-floor buses need to be able to attenuate higher-magnitude impulsive exposures. The new seat used on the low-floor bus in this study demonstrated that it was effective in attenuating these higher-magnitude impulsive exposures. Additional work with new seats should be conducted on high-floor buses to determine if new seats have the same attenuating effect, and further analyses should be done to determine whether there are frequency differences in the vibrational exposures between the two bus types.

References

ISO [1997]. Mechanical vibration and shock: evaluation of human exposure to whole-body vibration. Part 1: General requirements. Geneva, Switzerland: International Organization for Standardization. ISO 2631-1:1997.

ISO [2004]. Mechanical vibration and shock: evaluation of human exposure to whole-body vibration. Part 5: Method for evaluation of vibration containing multiple shocks. Geneva, Switzerland: International Organization for Standardization. ISO 2631-5:2004.

CHANGES IN SEAT-HEAD WHOLE-BODY VIBRATION TRANSMISSIBILITY AND MUSCLE ACTIVITY UNDER ASSYMETRIC NECK AND TRUNK POSTURES

Tammy Eger,[1] James P. Dickey,[2] Paul-Émile Boileau,[3] and Joan M. Stevenson[4]

[1]School of Human Kinetics, Laurentian University, Sudbury, Ontario, Canada
[2]Department of Human Health and Nutritional Sciences, University of Guelph, Ontario, Canada
[3]Institut de recherche Robert-Sauvé en santé et en sécurité du travail (IRSST), Montreal, Quebec, Canada
[4]School of Physical and Health Education, Queen's University, Kingston, Ontario, Canada

Introduction

Previous literature has shown that exposure to whole-body vibration (WBV) can lead to the development of low back pain and that nonneutral driving postures have also been linked to an increased risk of developing low back pain [Bovenzi and Betta 1994; Schwarze et al. 1998]. However, research that examines the combined effect of twisted sitting posture on trunk and neck muscle activity and the transmission of WBV up the spine is limited. This study set out to determine if seat-head vibration transmissibility and neck and trunk muscle activity were influenced by asymmetric trunk and neck postures.

Methods

Twelve volunteer male subjects with no previous history of low back or neck pain were exposed to vibration (0.1–20 Hz random noise Z-axis exposure at 0.8 or 1.2 m/s^2) while adopting one of six postures (no trunk rotation with 0°, 15°, or 45° neck rotation; 45° trunk rotation with 15° or 45° neck rotation; and maximal trunk and neck rotation). The postures were selected based on observed driving postures of LHD operators [Eger et al., in press]. The subjects sat on a rigid seat with no back rest. Muscle activity (EMG) was measured at six locations (right and left lower erector spinae (L3), right and left upper erector spinae (T9), left upper trapezius and the left sternocleidomastoid), and vibration transmissibility was measured between the rigid seatpan and head at the level of the occipital protuberance. Seat-head vibration transmissibility and coherence were monitored during each data collection trial with the Brüel & Kjær PULSE vibration measurement and analysis system. Every combination of vibration and posture condition was repeated twice, resulting in 24 randomized trials per subject. Each posture was held for 90 sec, with the first 15 sec without vibration and the last 75 sec with vibration. A custom software program was written in LabVIEW V7.1 (National Instruments, Austin, TX) and was used to process the EMG data and calculate the seat-head frequency response transfer functions. A general linear model repeated measures analysis of variance was used to determine statistical differences.

Results and Discussion

Posture had a significant effect on seat-head transmissibility magnitude (Figure 1), and there was a significant interaction between posture, vibration acceleration level, and the primary resonance frequency for seat-head transmissibility. There was a significant increase in muscle activity with an increase in vibration acceleration, and there was a significant interaction between vibration acceleration level, posture, and muscle activity in three of the four trunk (right L3, T9, and left T9) muscles monitored (Figure 2).

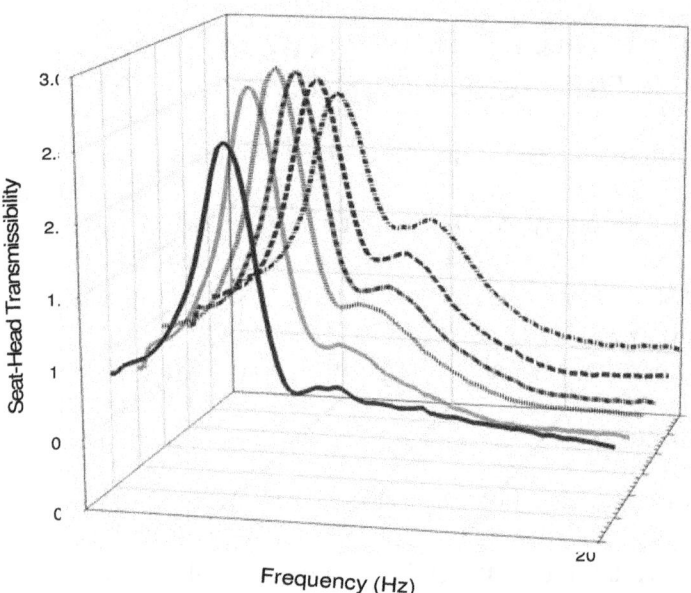

Figure 1.—Mean seat-head transmissibility (CSD transfer function) for each posture condition at 1.2 m/s² acceleration level. The postures are presented in order of increasing spinal and neck rotation.

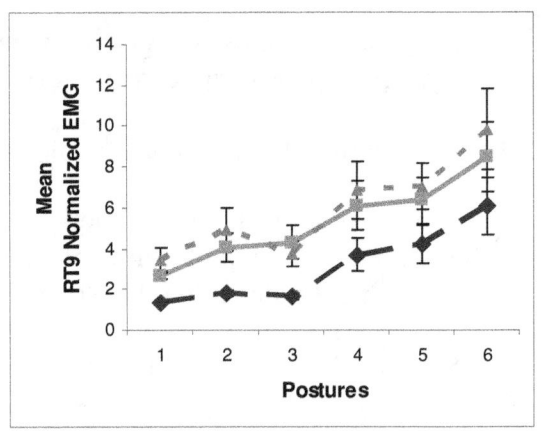

Figure 2.—Mean normalized EMG shown for erector spinae: right T9 level under no vibration exposure (diamonds and dashed line), 0.8 m/s² rms acceleration level (squares and solid line), and 1.2 m/s² rms acceleration level (triangles and dotted line). NOTE: The postures are presented along the horizontal axis in order of increasing trunk and neck rotation: P1(T0N0), P2(T0N15), P3(T0N45), P4(T15N45), P5(T45N45), P6(TmaxNmax).

Vibration is more readily transmitted through a rigid structure [Dietrich et al. 1991]; therefore, postures that increase the rigidity or stability of the spine will transmit vibration up the spine more readily [Mansfield et al. 2006]. The twisted trunk and neck postures (and resultant muscle activity increase) in the current study increased spinal stiffness. This likely accounts for the increase in transmissibility magnitude observed as the participants moved from the neutral posture to the maximally rotated trunk and neck posture. In conclusion, the findings from this study indicated that twisted neck and trunk postures influence the biomechanical response of the body to vibration.

Acknowledgment: Funding was provided by the Workplace Safety and Insurance Board of Ontario, Canada.

References

Bovenzi M, Betta A [1994]. Low-back disorders in agricultural tractor drivers exposed to whole-body vibration and postural stress. Appl Ergon 25(4):231–241.

Dietrich M, Kedzior K, Zagrajek T [1991] A biomechanical model of the human spinal system. J Eng Med 205:20–26.

Eger T, Stevenson J, Callaghan JP, Grenier S, VibRG [in press]. Predictions of health risks associated with the operation of load-haul-dump mining vehicles. Part 2: Evaluation of operator driving postures and associated postural loading. Int J Ind Ergon.

Mansfield N, Holmlund P, Lundström R, Lenzuni P, Nataletti P [2006] Effect of vibration magnitude, vibration spectrum and muscle tension on apparent mass and cross axis transfer functions during whole-body vibration exposure. J Biomech 39(16):3062–3070.

Schwarze S, Notbohm G, Dupuis H, Hartung E [1998]. Dose-response relationships between whole-body vibration and lumbar disk disease: a field study on 388 drivers of different vehicles. J Sound Vibration 215(4):613–628.

THE NATURE OF MULTIAXIS SIX-DEGREE-OF-FREEDOM VEHICLE VIBRATIONS IN FORESTRY, MINING, AND CONSTRUCTION HEAVY EQUIPMENT

James P. Dickey,[1] Tammy R. Eger,[2] Michele L. Oliver,[3] Paul-Émile Boileau,[4] and Sylvain Grenier[2]

[1]Department of Human Health and Nutritional Sciences, University of Guelph, Ontario, Canada
[2]School of Human Kinetics, Laurentian University, Sudbury, Ontario, Canada
[3]School of Engineering, University of Guelph, Ontario, Canada
[4]Institut de recherche Robert-Sauvé en santé et en sécurité du travail (IRSST), Montreal, Quebec, Canada

Introduction

Industrial exposure to whole-body vibration is associated with injury and discomfort. Certain industries, notably mining, construction, and forestry, involve high levels of vibration [Cann et al. 2003; Eger et al. 2006], including complex six-degree-of-freedom (df) vibration [Cation et al., in press]. Relatively few field studies have measured 6-df vibration exposures [Els 2005]. There is a general lack of awareness about the nature of multiaxis vibration exposures in heavy machinery. The purpose of this study was to evaluate the nature of multiaxis 6-df vehicle vibrations in heavy mobile equipment.

Methods

Measures of 6-df accelerations from the chassis were determined for load-haul-dump (LHD) vehicles, scrapers, and skidders from the Ontario mining, construction, and forestry industries. The data were collected during routine operations under actual working conditions. Unweighted chassis data were analyzed to determine the 6-df workplace vibration that is transmitted through seats to the workers.

Six-df vibration sensors were designed using a MEMSense MAG3D sensor. We tested a total of 11 LHDs (haulage capacity between 2.7 and 6.1 m^3), 13 scrapers, and 8 skidders (2 cable and 6 grapple). All data were sampled at 500 Hz and continuously recorded using a datalogger (DataLOG II, Biometrics) for approximately 1 hr per vehicle. The data were divided into successive 20-sec duration blocks, and the unweighted rms amplitude was calculated for each of the six vibration axes. Each of the 6 df was described by the mean and standard deviation. In order to evaluate the multiaxis nature of the vibrations, the 20-sec vibration exposures were divided into 33% percentiles (low, medium, and high) for each of the vibration axes. Each of the 20-sec vibration exposures was classified into 1 of the 729 possible combinations to describe the prevalence of combinations of multiaxis vibration.

Results

The vibration exposures varied among machine operations and type of machine, but overall were rather similar among industries (Figure 1). However, the nature of the multiaxis vibrations, as illustrated by the ranking of each of the industries, was considerably different across industries (Table 1). Although the "medium lateral acceleration, else low" condition was ranked fourth overall and occurred commonly in the forestry and construction datasets, it occurred infrequently in the mining dataset (ranked 149 of 729).

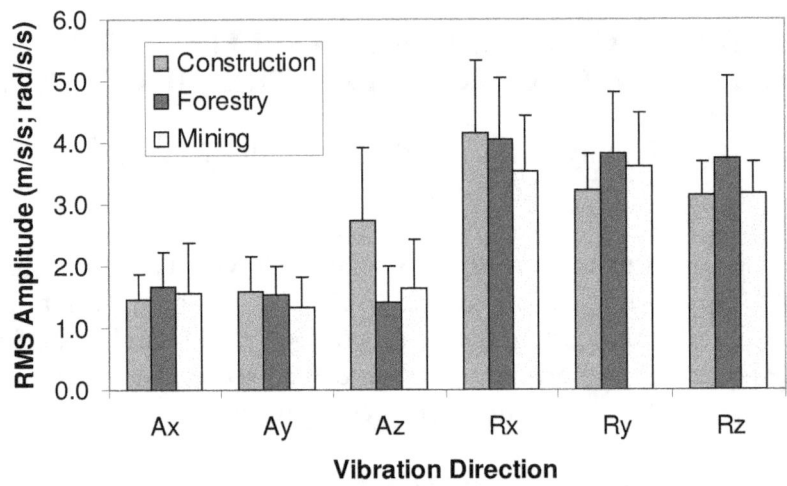

Figure 1.—Multiaxis vibration exposures during routine machine operation (rms amplitude, average, and standard deviation).

Table 1.—Four most prevalent combinations of vibrations in overall dataset and prevalence of these vibrations within each industry

Description	Ranking			
	Overall	Mining	Forestry	Construction
High on all 6 df	1	2	3	6
Low on all 6 df	2	20	2	1
Medium yaw, else low	3	1	14	35
Medium lateral, else low	4	149	1	22

Discussion

Although the vibration exposures of the individual vibration axes are similar among vehicles in the mining, construction, and forestry industries, the characteristics of the combined multiaxial vibration differ substantially. The magnitude and nature of these multiaxis vibrations are important for vehicle and seat design.

References

Cann AP, Salmoni AW, Vi P, Eger TR [2003]. An exploratory study of whole-body vibration exposure and dose while operating heavy equipment in the construction industry. Appl Occup Environ Hyg *18*(12):999–1005.

Cation S, Jack R, Oliver M, Dickey JP, Lee-Shee NK [in press]. Six degree of freedom whole-body vibration during forestry skidder operations. Int J Ind Ergon.

Eger T, Salmoni A, Cann A, Jack R [2006]. Whole-body vibration exposure experienced by mining equipment operators. Occup Ergon *6*(3/4):121–127.

Els PS [2005]. The applicability of ride comfort standards to off-road vehicles. J Terramechanics *42*(1):47–64.

This project was funded by a grant from the Workplace Safety and Insurance Board (Ontario, Canada).

Poster Session

Chair: Bertrand Valero
Co-Chair: Kimberly Balogh

Presenter	Title	Page
A. Øvrum * University Hospital of North Norway	Whole-Body Exposure from Heavy Loading Vehicles with Different Risk Assessment Outcomes Using Recommendations in the ISO-2631 and ISO-8041 Standards	50
B. Valero * University of Illinois at Chicago	Ride Comfort Evaluation Using a Three-Dimensional Model of a Human Body Focusing on the Lumbar Spine Area	65
J. Kim * University of Cincinnati	Development of a Receptance Method-Based Modeling Technique for Finite-Element Analysis of Hand-Arm Vibration	72
D. Perottino ** Polytechnic University of Turin	An Insight Into Using the Tire as an Active Suspension in Human Body Vibration Control	122
R. E. Larson ** Exponent, Inc.	Measurement and Evaluation of Vibration Exposure for a Kawasaki 80Z III Wheeled Loader	124
I. M. Dudnyk ** Donetsk National Medical University	Vibration as a Risk Factor and the Need to Prevent Vibration Disease Among Miners in the Donets Coal Basin	126
J. M. Hughes ** Johns Hopkins University	Increased Oxidant Production by Endothelial Nitric Oxide Synthase in a New Rodent Model of Hand-Arm Vibration Syndrome	128
I. J. H. Tiemessen University of Amsterdam	Development of an Intervention Program to Reduce Whole-Body Vibration Exposure at Work Induced by Change in Behavior: A Study Protocol	130
M. L. M. Duarte ** Federal University of Minas Gerais	Whole-Body Vibration Exposure Values for Car Passengers on Rough Roads: A Focus on Health	132
Tammy Eger School of Human Kinetics	ISO 2631-1 and ISO 2631-5: A Comparison of Predicted Health Risks for Operators of Load-Haul-Dump Vehicles	134

* Also given as oral presentation.
** Author unable to attend conference.

AN INSIGHT INTO USING THE TIRE AS AN ACTIVE SUSPENSION IN HUMAN BODY VIBRATION CONTROL

Daniele Perottino,[1] Farid Amirouche,[2] Bertrand Valero,[2] and Eurelio Soma[1]

[1]Polytechnic University of Turin, Italy
[2]Vehicle Technology Laboratory, University of Illinois at Chicago

Introduction

Earthmoving equipment relies greatly on tires to move heavy loads and work on rugged terrain. Tires have for years been the focus point of many analyses, mostly in relation to the safety of the vehicles. The vibration transmitted to the operator has in many ways been neglected or simply secondary to tire design. Long-term exposure to vibration has been linked for years to severe low back pain, especially in the lumbar region [Schwarze et al. 1998]. The primary suspension of most heavy vehicles is the tire. In the case of earthmoving vehicles such as telehandlers, the tire pressure and size are primary factors in the transfer of energy to the cab and seat. Thus, the only component limiting vibration to the operator is the tire.

The focus of the present collaborative study between the Biomechanics Research Laboratory at the University of Illinois at Chicago (UIC) and the Polytechnic University of Turin, Italy, is to characterize the material and inertia properties of a typical telescopic vehicle tire in order to create a dynamic model of the tire to investigate how design parameters can be controlled during a particular task or maneuver. The work combines both the tire as a finite-element dynamic model (flexible body) and an intricate human model that can be used to assess the spinal response to this tire/suspension mechanism. This research seeks to provide an insight into the tire as a primary suspension for vibration control and to evaluate its role in safety and comfort of the operator.

Methods

A 240- by 120-mm grid was traced on one of the four tires of a telescopic vehicle. A load was then applied to the forklift. The load was slowly decreased. The deformation of the grid was observed. The same protocol was followed for different pressures inside the tire.

The experimental data collected were used to validate a finite-element model of the tire. The static model includes the geometrical and material nonlinearities of the tire. The finite-element model was then exported to ADAMS (Automatic Dynamic Analysis of Mechanical Systems). A revolute joint was placed at the center of the rim to allow the rotation of the tire. The contact between the ground and the tire was made through the contact joint. The ground was modeled as a plane element. A contact translational velocity of 20 km/hr was imposed to the plane. The displacements of the center of mass were collected for the different pressures considered.

The displacements resulting from the dynamic analysis were applied to the seat of a sitting human body model developed in the UIC Biomechanics Lab. The assumption of a direct connection between the displacement of the tire center of mass and the displacement of the seat was made. The human body model includes 19 segments and 18 joints. The inertia properties for each segment were taken from an anthropometric database. The interaction between the body and seat was modeled through spring and damper in parallel, according to the Amirouche et al. [1997] model. The characteristics of each joint were tuned in order to match experimental data found in the literature about the first human body resonance. A finite-element model of the lumbar spine developed at the UIC Biomechanics Lab was used for the combined analysis.

Results

The results showed lower displacement of the operator's neck when tire pressure is decreased. Essentially, a softer tire leads to better comfort if the pressure is adjusted as a function of stresses measured at the rim. Thus, a tire can act as an active suspension if pressure can be controlled within a certain bandwidth.

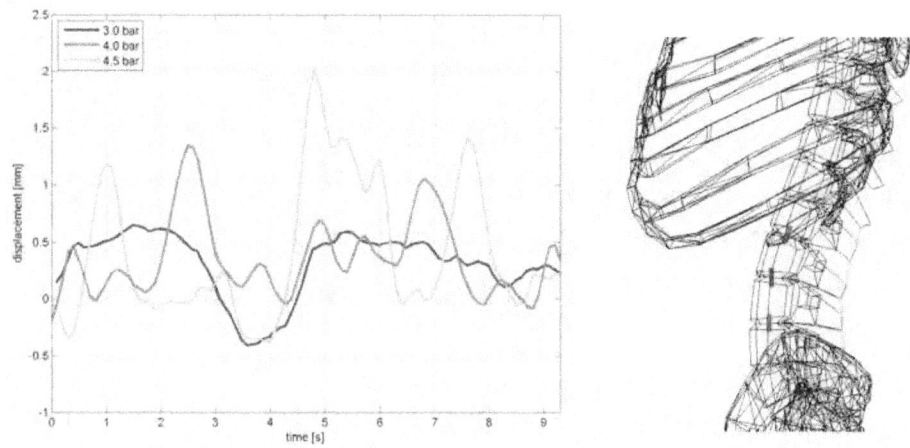

Figure 1.—Human body model and graph of the neck displacement versus time for different pressures.

Figure 2.—Dynamic tire model *(left)* and telescopic vehicle *(right)*.

References

Amirouche F, Alexa E, Xu P [1997]. Evaluation of dynamic seat comfort and driver's fatigue. Warrendale, PA: Society of Automotive Engineers, Inc., technical paper 971573.

Schwarze S, Notbohm G, Dupuis H, Hartung E [1998]. Dose-response relationships between whole-body vibration and lumbar disk disease: a field study on 388 drivers of different vehicles. J Sound Vibration *215*(4):613–628.

MEASUREMENT AND EVALUATION OF VIBRATION EXPOSURE FOR A KAWASAKI 80Z III WHEELED LOADER

Robert E. Larson and Dana Hansen
Exponent, Inc., Phoenix, AZ

Introduction

Measurements of the vibration and impact environment when operating a Kawasaki 80Z III wheeled loader were made at the Canadian Pacific Classification Yard, Bensenville, IL, in January 2004. The health implications of exposure to the measured loader vibration levels were evaluated by comparison with the human vibration exposure boundaries given in the applicable standards and measurements of peak acceleration levels associated with many common daily activities such as walking, climbing stairs, and jogging.

Methods

To characterize the vibration exposure that a loader operator would experience, the loader was instrumented to record acceleration on the operator's seat and on the floor directly under the seat, per ISO standard 2631-1 (1997). The loader was equipped with an air suspension seat and ride control feature, although the ride control feature was turned off during all of the testing. The speed of the loader was measured with a global positioning system speed sensor to characterize the operations being recorded. The loader did not have a bucket attached; rather, it was being used for functions such as hauling rail ties and thus had a fork attachment. The loader was driven around the yard and used to perform hauling and transit operations at speeds up to 8–9 mph. These particular maneuvers were performed to record the vibration level under what was described as typical operation in this yard. In addition to this typical operation, measurements were also made to characterize some of the extreme conditions that an operator might occasionally experience. This included traveling back and forth across a series of rails, traveling parallel to the track with one side's wheels between the rails, and traveling across a series of railroad ties lined up on a roadway. Four categories of activities were evaluated: "driving," "cross rails," "over ties," and "tire between rail." The individual measurements within each category were combined for overall exposure level calculations. It should be noted that the results presented here do not represent the exposure for wheeled loader activities such as digging with a bucket attachment or high-speed operation over rough terrain.

Results

The overall weighted root-mean-square (rms) acceleration (A_w) for each of the four activities is shown in Figure 1. A guide to interpreting weighted acceleration values with respect to health is given in Annex B of ISO 2631-1. A health guidance caution zone (HGCZ) is defined to indicate the level of vibration where a health risk could exist. The amount of time for reaching the HGCZ for the "driving" runs is 10.6 hr in the direction of vertical vibration, the axis with the highest levels of vibration. Using ISO 2631-1 based on an estimated vibration dose value (VDV), the time for the "driving" runs to reach the health caution zone is 24.3 hr. Although the "driving" runs best represent the exposure that an operator at a rail yard might experience over an extended period of time, the cumulative exposure while traversing "cross rails" was also examined. In this case, the vibration in the lateral direction was the highest, resulting in durations of 3.5–4.0 hr before reaching the lower boundary (depending on which time-dependency equation is used).

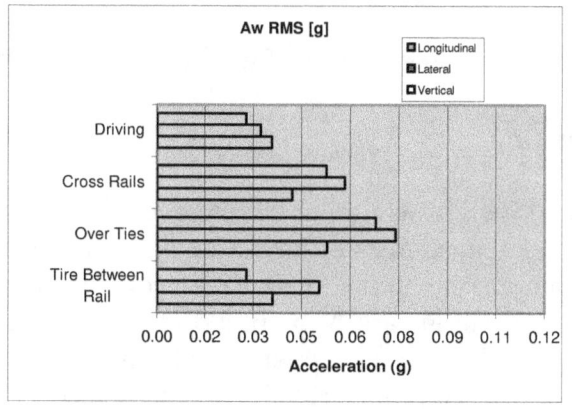

Figure 1.—Weighted rms acceleration.

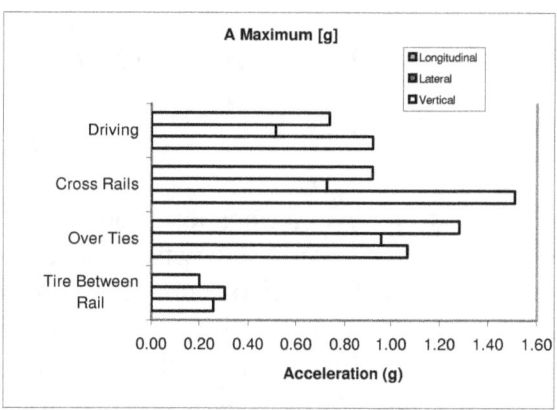

Figure 2.—Peak acceleration.

To evaluate the effect of transient vibration and shock, a VDV was calculated for the "driving" runs. The VDVs were 0.38, 0.48, and 0.54 $g \cdot s^{0.25}$ in the longitudinal, lateral, and vertical directions, respectively. These values are well below the action level of 1.53 $g \cdot s^{0.25}$ defined in British Standard 6841 (1987), the EU Directive action value of 0.93 $g \cdot s^{0.25}$, and the EU exposure limit of 2.14 $g \cdot s^{0.25}$. The "driving" runs were also evaluated using ISO standard 2631-5 for evaluating vibration containing multiple shocks. The resulting values, based on a lifetime of occupational exposure, were below the thresholds, indicating that there is low probability of an adverse health effect.

The rail-crossing runs, traversing railroad ties, and traveling with one side between the rails represent occasional exposures to isolated impact and vibration events. Therefore, it is worthwhile to look at the peak values from these runs and consider the magnitude of these events rather than the weighted average values. As seen in Figure 2, the highest acceleration value recorded was in the vertical direction while crossing rails (1.51 g). The highest longitudinal and lateral values were found while traversing the railroad ties (1.28 and 0.96 g, respectively). To put the magnitude of these acceleration values in perspective, they can be compared with values that have been measured for common daily activities. The acceleration peaks at the base of the lumbar spine associated with walking down stairs or walking fast were found to be approximately 1.0 g, stepping down a 9-in step was at 1.6–2.2 g, and jogging was found to be in the range of 1.9–3.9 g. It is clear that many common daily activities create higher peak vertical accelerations in the lumbar spine than the occasional vertical transients experienced while operating a loader.

Discussion

The vibration exposure experienced while operating a Kawasaki loader in a rail yard was found to be well below the HGCZs for continuous activity. The transient and peak levels also did not seem to be at a level of concern. The direction of highest acceleration was dependent on what activity the loader was being used for, although vertical was highest for transit and hauling activities and for unweighted peak acceleration when crossing rail.

VIBRATION AS A RISK FACTOR AND THE NEED TO PREVENT VIBRATION DISEASE AMONG MINERS IN THE DONETS COAL BASIN

Igor M. Dudnyk
Donetsk National Medical University, Donetsk, Ukraine

Introduction

The extraction and transportation of coal and rock from a mine are achieved with a variety of machinery and processes, which constitute inherent sources of vibration and mechanical shock. The effects of a vibratory mining environment on the human body have been documented and can lead to occupational illness or vibration disease in miners. Vibration disease is the second leading cause of occupational illness in miners working in the Donets coal basin (Donbass) in the Ukraine. One component of the disease, hand-arm vibration (HAV), is experienced by miners who extract coal at the working face using picks and by drifters who perform boring and blasting activities associated with entry development. Operators of underground locomotives experience symptoms associated with the other component of vibration disease—whole-body vibration (WBV).

Methods

Incidences of HAV and WBV in underground mines were investigated among face workers, drifters, and locomotive operators. The procedures for measuring and analyzing acceleration data were conducted in accordance with national and international standards. An index from Suvorov et al. [1984] was used to obtain the total vibration dosage experienced by these workers.

Results

Exposure levels were measured at the handles of jackhammers used by the miners. Maximum vibration levels of 2.0–2.1 m/s^2 (93–98 dB) were recorded at frequencies from 16 to 63 Hz. The occupational exposure limit (OEL) designated for the low- to mid-vibration frequencies is the most often exceeded, whereas vibration frequencies in the range of 250–1000 Hz that are measured for jackhammers generally do not exceed the OEL.

The vibration characteristics of two types of jackhammers used by miners were compared relative to HAV exposure levels. The MP-2 jackhammer indicated worse results in the low- and mid-frequency ranges than a second jackhammer type. However, HAV levels were approximately the same at frequencies of 63 and 125 Hz. Drifters are exposed to HAV as they drill blast holes and load material during coal excavating operations. Drilling is performed using perforators (punchers – model PP54) that are equipped with pneumatic supports. Exposure duration from these sources of HAV averages no less than 2 hr per place change. Maximum HAV levels were recorded for frequencies from 8 to 31.5 Hz. An exposure level of 1.8 m/s^2 (84 dB), exceeding the OEL, occurred at a frequency of 16 Hz.

Besides HAV, drifters (miners loading rock from a rail-mounted loading machine) are exposed to WBV. During these duties, the drifter operates the loading machine from a step or platform on the machine. Maximum levels of vibration in the frequency range of 2–16 Hz were shown to exceed the OEL. Duration for completing this activity is no more than 1 hr. The drifter will occasionally operate the loading machine from ground level (off the machine) to perform work in narrow developments and short distances from the terminal end of the railway up to the working face.

Locomotive operators experience WBV as they transport coal and rock out of the mine workings via the rail system. Measurements of WBV exposure indicate that the OEL is exceeded at certain exposure frequencies. Measurements taken on the floor of the operator deck of an electric locomotive (model AM-8) show the exposure limit is exceeded by 0.04 m/s^2 (2 dB) at a frequency of 2 Hz and by 0.09 m/s^2 (4 dB) at a frequency of 4 Hz. Maximum exposure levels were recorded for operators of underground electric locomotives that show the OEL is exceeded by 0.04 m/s^2 (2 dB) and 0.09 m/s^2 (4 dB) for frequencies of 8 Hz and 16 Hz, respectively. Total exposure duration for operators of both types of locomotives is approximately the same and varies up to 4 hr for area changes.

When analyzing the vibration exposure characteristics of the mine machinery investigated, one can draw some general conclusions about the nature of vibration exposures for the mining job activities cited above. Devices such as jackhammers and perforators demonstrate a vibration described as variable impulsive, as opposed to variable non-impulsive vibration, which is associated with borehole drilling. Considering the coal strata characteristics indigenous to the central Donbass, it is not necessary for the jackhammer to operate continuously at high force levels associated with peak power. Typical jackhammer operation shows forces ranging from 98 to 147 N. Consequently, HAV exposures influenced by operating the jackhammer at peak force levels are not significant. Noteworthy aspects of using the perforator are greater energy and frequency of striker impact (1.5 times that of a jackhammer) and the presence of a twisting moment during the active period of its operation.

An element of rail transport and travel includes mechanical vibrations, which are transmitted to locomotive operators through the vehicle seat and floor. During this study, measurements of WBV exposure show potentially harmful levels for miners engaged in coal production activities. These circumstances require regulatory oversight that incorporates protective measures for mine face workers, drifters, and locomotive operators in order to lower the risks of HAV and WBV exposures that can lead to occupational illness or vibration disease among underground miners.

Concluding Remarks

Considering the vibratory work environment and associated risk factors for mine face workers, the work activities of the drifter pose the highest risk of vibration disease. Thus, it is important for mining operations to include a system of preventive measures to lower the risk of vibration disease among mine face workers, particularly the drifter.

Reference

Suvorov GA, Shkarinov LN, Denissov EI [1984]. Hygienic assessment of occupational noises and vibrations (in Russian). Meditsina (Moscow):57–59.

INCREASED OXIDANT PRODUCTION BY ENDOTHELIAL NITRIC OXIDE SYNTHASE IN A NEW RODENT MODEL OF HAND-ARM VIBRATION SYNDROME

Jennifer M. Hughes,[1] Oliver Wirth,[2] Kristine Krajnak,[2] Sheila Flavahan,[1] G. Roger Miller,[2] and Nicholas A. Flavahan[1]

[1]Department of Anesthesiology, School of Medicine, Johns Hopkins University, Baltimore, MD
[2]Health Effects Laboratory Division, National Institute for Occupational Safety and Health, Morgantown, WV

Introduction

The predominant vascular symptom of hand-arm vibration syndrome (HAVS) is exaggerated vasoconstriction, or Raynaud's phenomenon, involving the digital arteries. In a rat tail model of HAVS, we previously demonstrated that a single exposure to vibration caused a persistent and selective increase in smooth muscle vasoconstriction in the tail artery mediated by activation of α2C-adrenoceptors [Krajnak et al. 2006]. The goal of the present study was to analyze the effects of vibration on digital arteries using a new rat model of HAVS.

Methods

Paws of rats were exposed to a single period of vibration (4 hr, 125 Hz, constant acceleration 49 m/s^2 root-mean-square) using a restraint method that maintains contact between the rat paw and a vibrating platform (Figure 1). The physical or biodynamic response of the paw in this preparation was assessed previously using a laser vibrometer. Results showed amplification of vibration transmissibility to the paw at this frequency compared with vibration at the platform. After the rats were euthanized, digital arteries were carefully dissected from vibrated and nonvibrated control paws and cannulated at both ends with glass micropipettes in a microperfusion

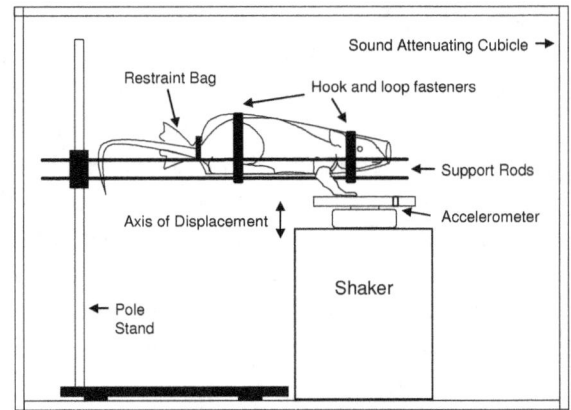

Figure 1.—Exposure apparatus showing rat restrained in a cone-shaped plastic bag and the paw contacting the vibrating platform.

chamber. The chamber was superfused with modified Krebs-Ringer bicarbonate solution, maintained at 37 °C, pH 7.4, and the artery segment was pressurized to a constant transmural pressure of 60 mm Hg. The vessel image was projected onto a video monitor, and the internal diameter was continuously measured using a video dimension analyzer (Living Systems Instrumentation, Burlington, VT) and recorded using a BIOPAC data acquisition system (Santa Barbara, CA). Cumulative concentration-effect curves to the α1-adrenoceptor agonist phenylephrine (0.001–1 µM) or 5-hydroxytryptamine (5-HT; 0.001–0.03 µM) were generated by increasing agonist concentration in half-log increments after the response to the previous concentration had stabilized. Cumulative concentration-response curves to acetylcholine were assessed during a stable constriction to phenylephrine (by 30% of baseline). In some arteries, the endothelium was mechanically removed, taking care to not damage the underlying smooth muscle. The production of nitric oxide (NO) in the digital arteries was determined using the NO-sensitive fluorescent probe 4-amino-5-methylamino-2′,7′-difluorofluorescein diacetate (DAF-FM DA). The production of reactive oxygen species (ROS) was determined using the H_2O_2-sensitive fluorescent probe 5- (and 6-)chloromethyl-2′,7′-dichlorodihydrofluorescein diacetate (both at 5 µM). Images were captured using a fluorescent microscope every 10 sec after addition of fluorescent probes and compared to baseline levels.

Results

In vitro analysis of vascular function demonstrated that constriction of digital arteries to phenylephrine or 5-HT was blunted in vibrated arteries ($p < 0.05$). $\alpha2$-adrenoreceptor activation by the agonist UK14304 did not cause significant constriction of these arteries. Relaxation to the endothelium-dependent dilator acetylcholine was not significantly different between control and vibrated arteries. Endothelium removal abolished relaxation to acetylcholine. After endothelium denudation, vasoconstriction to phenylephrine or to 5-HT was no longer impaired in vibrated arteries. Likewise, the inhibitor of endothelial NO synthase (NOS) N^G-nitro-L-arginine methyl ester (L-NAME; 100 µM) increased vasoconstriction to phenylephrine or to 5-HT in vibrated but not control arteries ($p < 0.05$) and abolished the vibration-induced depression in vasoconstrictor responses. However, the production of NO, as determined using the NO-sensitive probe DAF-FM DA was actually reduced in vibrated compared with control arteries ($p < 0.05$). Concomitantly, ROS production, as determined using the H_2O_2-sensitive probe, was increased in vibrated compared with control arteries ($p < 0.05$). This vibration-induced increase in ROS production was abolished by the NOS inhibitor L-NAME (100 µM) or by the H_2O_2 inhibitor catalase (1,000 units/mL). Catalase (1,000 units/mL) also increased vasoconstrictor responses to phenylephrine or 5-HT in vibrated but not control arteries and abolished the vibration-induced depression in vasoconstrictor responses.

Discussion

The major findings of the present study were the following: (1) Digital arteries from vibrated rat paws demonstrate blunted vasoconstrictor reactivity when compared to controls; (2) endothelial denudation or inhibition of endothelial NOS equalized vasoconstrictor responses between control and vibrated arteries; (3) vibrated arteries had decreased NO production, but increased ROS production compared to control arteries; and (4) the vibration-induced increase in ROS production was abolished by inhibition of NOS or by extracellular catalase, which also abolished the vibration-induced depression in vasoconstrictor responses. The results suggest that vibration causes an uncoupling of endothelial NOS, which promotes the generation of ROS rather than NO [Forstermann and Munzel 2006]. The endothelial and NOS-dependent generation of ROS results in oxidant stress and diffusion of endothelium-derived H_2O_2 through the blood vessel wall, which depresses vasoconstrictor responses. This early event may be a key step in the pathogenesis of HAVS.

References

Förstermann U, Münzel T [2006]. Endothelial nitric oxide synthase in vascular disease: from marvel to menace. In: Circulation *113*(13):1708–1714.

Krajnak K, Dong RG, Flavahan S, Welcome D, Flavahan NA [2006]. Acute vibration increases α_{2C}-adrenergic smooth muscle constriction and alters thermosensitivity of cutaneous arteries. J Appl Physiol *100*(4):1230–1237.

DEVELOPMENT OF AN INTERVENTION PROGRAM TO REDUCE WHOLE-BODY VIBRATION EXPOSURE AT WORK INDUCED BY CHANGE IN BEHAVIOR: A STUDY PROTOCOL

Ivo J. H. Tiemessen, Carel T. J. Hulshof, and Monique H. W. Frings-Dresen

Coronel Institute for Occupational Health, Academic Medical Center, University of Amsterdam, Amsterdam, the Netherlands

Introduction

It has been estimated that 4%–7% of all drivers in some European countries, the United States, and Canada are exposed to potentially harmful whole-body vibration (WBV). Long-term occupational exposure to WBV is associated with an increased risk of low back pain and disorders of the lumbar spine [Lötters et al. 2003]. Despite this, a recent review revealed that literature on successful strategies to reduce WBV exposure in the workplace is scarce [Tiemessen et al. 2007]. This review distinguished strategies directed toward (1) (technical) design considerations and (2) skills and behavior. The emphasis in recent research has been on the first category, although results (especially based on behavioral changes) from the second category may be more promising in reducing WBV exposure at work. Therefore, we developed an intervention program aimed at reducing WBV exposure in a population of drivers with an emphasis on changing the behavior of the driver and the employer. WBV exposure is used as a proxy for low back complaints.

Methods

We developed an intervention program to change the behavioral aspects toward WBV exposure in accordance with the ASE model [De Vries et al. 1998], which is often used in the field of health education. This ASE model can be used to explain attitude and attitudinal changes. The ASE model differentiates three cognitions: attitude, social influence, and self-efficacy. Our intervention program focused on (1) a participative approach with a role for the driver, the employer, and the occupational health professional, and (2) a multifaceted approach comprising both design considerations and skills and behavior aimed at WBV reduction, with emphasis on the latter category. By increasing knowledge and skills toward changing these determinants, the ASE of both drivers and employers will be positively changed, which in turn will have a favorable effect on the level of exposure.

The WBV exposure of the drivers and the process variables knowledge, attitude, and behavior both for the drivers and the employers were assessed with field vibration measurements according to ISO 2631-1 and by self-administered questionnaires. Process variables, such as the intention to change behavior, impediments in implementing the intervention program, and compliance with the intervention program, were also assessed by the questionnaire.

Discussion

This study entailed the development of an intervention program aimed at reducing WBV exposure to drivers at work by inducing a change in behavior or at least an intent to change the behavior of the drivers and their employers. The program consisted of (1) an individual health surveillance, (2) an informational brochure, (3) an informative oral presentation, and (4) a report of the results of the performed measurements of the WBV magnitude. The "care as usual" group received care as usual, which consisted only of the informational brochure and the report.

This type of intervention program has several advantages. First, it may lead to more permanent effects (until the behavior is changed again). This long-term effect might work simultaneously with changes in WBV company policy. Another advantage is that WBV reduction may be realized without large investments in money or time. Third, by implementing this type of intervention program, not only the drivers bear responsibility for a possible reduction in WBV exposure, but the employers also have a specific role because the program implies certain changes in WBV company policy. Lastly, this intervention program provides an active, participative approach of implementing preventive measures. This is in accordance with a study by van der Molen [2006], who concluded that in the field of occupational health and safety, actively implementing preventive measures with a participative component is an important facet of the success of an intervention program.

The near future will show if this intervention program is effective by showing a decrease in WBV exposure. A change in attitude or the intention to change one's attitude toward WBV exposure will hopefully contribute to a decrease in WBV exposure for drivers of various vehicles and low back pain over the long term.

References

De Vries H, Mudde AN, Dijkstra A, Willemsen MC [1998]. Differential beliefs, perceived social influences, and self-efficacy expectations among smokers in various motivational phases. Prev Med *27*(5):681–689.

Lötters F, Burdorf A, Kuiper J, Miedema H [2003]. Model for the work-relatedness of low-back pain. Scand J Work Environ Health *29*:431–40.

Tiemessen IJH, Hulshof CTJ, Frings-Dresen MHW [2007]. An overview of strategies to reduce whole-body vibration exposure on drivers: a systematic review. Int J Ind Ergon *37*(3): 245–256.

van der Molen HF [2006]. Evidence-based implementation of ergonomic measures in construction work [Thesis]. Amsterdam, Netherlands: University of Amsterdam, pp. 165–176.

WHOLE-BODY VIBRATION EXPOSURE VALUES FOR CAR PASSENGERS ON ROUGH ROADS: A FOCUS ON HEALTH

M. L. M. Duarte, E. A. Oliveira, and L. V. Donadon

Federal University of Minas Gerais, Mechanical Engineering Department,
Group of Acoustics and Vibration in Human Beings, Belo Horizonte/MG, Brazil

Introduction

It is important for the safety and comfort of drivers to understand the effects of whole-body vibration (WBV) on their everyday driving activities. The goal of this study was to evaluate, from a health perspective, WBV levels on standard passenger car drivers when driving their vehicles over rough road. Some studies have reported on the effects of WBV on the human spine [Bovenzi and Hulshof 1999] and on the comfort and health of drivers [Balbinot 2001]. However, there is no evaluation procedure (for example, WBV standards) that can adequately determine the effects of WBV exposure on humans [Griffin 1990].

Methods

This laboratory study examined the influence of WBV on the health of typical passenger car drivers. The test sample included students who drove a test route featuring different roadway surfaces on the university campus. The vehicles investigated differed by make and model. The tests were performed according to European Directive 2002/44/EC [2002] (based on ISO 2631-1 (1997)), which established the exposure action value (EAV) and exposure limit value (ELV) for assessing an 8-hr vibration exposure period.

The test sample of subjects was composed of 12 volunteers—11 students and 1 lecturer from the Mechanical Engineering Department of the Federal University of Minas Gerais. Fourteen tests were performed with 14 different vehicles. Two volunteers took part twice in the experiments and each drove two different vehicles. Of the different vehicles used, one can be described as off-road; the others were regular passenger cars. The vehicles were considered representative of those commonly driven in Brazil. No particular vehicle make or model was favored over another. Most of the vehicles were considered "compact," with small-sized engines. Information describing both the drivers and the vehicles was recorded for each test. Descriptive characteristics for drivers included sex, age, body mass index, and subjective comfort levels obtained at the end of the test. Vehicle information included the make, model, production year, engine size, suspension (front and rear), transmission (number of gears), tires, number of doors, and maximum occupancy. The study results were viewed with regard to other potential influencing factors, such as number of vehicle passengers, the presence of heavy objects inside the vehicle, noise levels, and vehicle condition (i.e., how well maintained) [Oliveira 2006].

The test route included two types of road surfaces (asphalt and parallelepipeds[2]) with speed restraints (or bumps). The route was considered more severe or "rougher" than normal pavements found in the urban community. This allowed for extrapolating the results to less severe road surface conditions. In selecting test route conditions, researchers attempted to represent the normal driving routes common to all test subject drivers. Moreover, this permitted measuring WBV exposure levels that subjects experience on a daily basis. Drivers were not

---

[2]Parallelepipeds are stones used for paving roads and are so called because of their shape. They are very common in small cities and small traffic roads in Brazil as a cheaper way of paving roads.

instructed as to how to drive the test vehicles, except that they operate them in a manner similar to their normal driving habits. The tests were always performed in the evening, between 7:00 p.m. and 9:00 p.m., so that outside temperature did not influence the results. The total driving distance for each of the tests was about 3.6 km and lasted a total of 7 min.

Results

Weighted root-mean-square accelerations were obtained from measurements in all three orthogonal directions using a triaxial accelerometer and a portable analyzer (model Maestro from 01dB), which included the weighting functions of the ISO 2631-1 (1997) standard. The results showed WBV exposure levels in the z direction as the dominant (or worse-case) axis for most of the vehicles. The next highest exposure levels were in the x direction. Two vehicles showed the y direction as the dominant (or worse-case) axis, and thus WBV exposures were higher than in the other (z and x) directions. Moreover, the y-axis exposure levels for these two vehicles exceeded the EAV set by European Directive 2002/44/EC [2002]. Nevertheless, none of the vehicles exceeded the ELV recommended for an 8-hr exposure period.

Discussion

Although the evaluations were performed for 7-min test periods, the results could be applied to an 8-hr equivalent exposure period. In this regard, the cars show satisfactory vibration isolation for the drivers. For some of the vehicles, however, the results show that the EAV would be achieved in less than a 30-min drive. The present study may thus be used as a basis for improving vehicle dynamics and seats in passenger cars typically driven in Brazil.

Acknowledgment: This work was developed using funds from Fundação de Amparo à Pesquisa do Estado de Minas Gerais (FAPEMIG) under grant TEC932–05.

References

Balbinot A [2001]. Characterization of vibration levels in bus drivers: a focus on comfort and health (Thesis) (in Portuguese). Porto Alegre, Brazil: Federal University of Rio Grande do Sul.

Bovenzi M, Hulshof CT [1999]. An updated review of epidemiologic studies on the relationship between exposure to whole-body vibration and low back pain (1986–1997). Int Arch Occup Environ Health 72(6):351–365.

Directive 2002/44/EC [2002]. Directive 2002/44/EC of the European Parliament and of the Council of 25 June 2002 on the minimum health and safety requirements regarding the exposure of workers to the risks arising from physical agents (vibration).

Griffin MJ [1990]. Handbook of human vibration. London: Academic Press.

ISO [1997]. Mechanical vibration and shock: evaluation of human exposure to whole-body vibration. Part 1: General requirements. Geneva, Switzerland: International Organization for Standardization. ISO 2631-1:1997.

Oliveira EA [2006]. Whole-body vibration effects caused by automotive vehicles in human beings: a focus on comfort and health (in Portuguese). Belo Horizonte/MG, Brazil: Federal University of Minas Gerais, Mechanical Engineering Department.

ISO 2631-1 AND ISO 2631-5: A COMPARISON OF PREDICTED HEALTH RISKS FOR OPERATORS OF LOAD-HAUL-DUMP VEHICLES

Tammy Eger,[1] James P. Dickey,[2] and Sylvain G. Grenier[1]

[1]School of Human Kinetics, Laurentian University, Sudbury, Ontario, Canada
[2]Department of Human Health and Nutritional Sciences, University of Guelph, Ontario, Canada

Introduction

Operators of load-haul-dump (LHD) mining vehicles are exposed to whole-body vibration (WBV) and shocks during the course of their work [Eger et al. 2006]. ISO standard 2631-1 (1997) provides guidance on quantifying WBV in relation to human health and comfort. In 2004, ISO 2631-5 was established to quantify health effects to the lumbar spine and the vertebral endplates based on seated WBV exposure containing multiple shocks. Few studies have reported predicted health risks based on ISO 2631-1 and 2631-5 standards. Johanning et al. [2006] and Cooperrider and Gordon [2006] reported low health risks (for locomotive operators) based on ISO 2631-5 criteria and higher health risks based on ISO 2631-1 criteria. However, Alem [2005] reported higher predicted health risks (for army vehicle operators) according to ISO 2631-5 criteria compared with those based on ISO 2631-1 criteria. Therefore, health risks predicted by ISO 2631-1 and 2631-5 were compared to determine if the two approaches predicted similar health risks for LHD operators.

Methods

WBV exposure at the operator/seat interface from 19 LHD vehicles were recorded for a minimum of 1 hr (and extrapolated to represent exposure typical of an 8-hr work shift). Field measurements of WBV were conducted in accordance with the ISO 2631-1 standard. Health risks predicted by the ISO 2631-1 and ISO 2631-5 analysis were compared across the common values that could be normalized to an 8-hr equivalent value. Rankings (high, moderate, low) for predicted health risks were assigned to each LHD vehicle operator based on health guidance caution zone (HGCZ) limits for A(8) and VDV_{total} values published in ISO 2631-1 and S_{ed} limits and R Factor values established for probability of an adverse health effect as reported in ISO 2631-5.

Results

According to criteria established in the ISO 2631-1 standard, 9 LHD vehicle operators were exposed to vibration levels above the HGCZ and 10 were exposed to WBV levels within the HGCZ (Figure 1). According to criteria established in ISO 2631-5, 2 LHD operators were exposed to vibration that placed them in a category associated with a "high probability" of injury to the lumbar spine, 2 were predicted to have a moderate injury risk, and the remaining 15 were predicted to have a low probability of injury (Figure 1).

Discussion

The ISO 2631-5 standard was established to evaluate vibration exposures with high shock content, and vibration signals with crest factor values greater than 9 indicate high shock content within a vibration signal. In this study, all the averaged crest factor values recorded in the vertical axis were above 9, and the highest recorded value was over 50. Given these results, it would be logical to conclude that LHD operators experience vibration levels with high shock content, making an ISO 2631-5 analysis important.

Health risks predicted by the ISO 2631-5 standard were generally lower than those predicted by the ISO 2631-1 standard. The boundaries for probable health effects reported in ISO 2631-5 were established using the best guidance at the time; however, they have not been epidemiologically validated. This study provides evidence to suggest that the ISO 2631-5 boundaries for probable health effects could be set too high. Further research with other populations is required to determine if changes to the standards are required.

Acknowledgment: Funding was provided by the Workplace Safety and Insurance Board of Ontario, Canada.

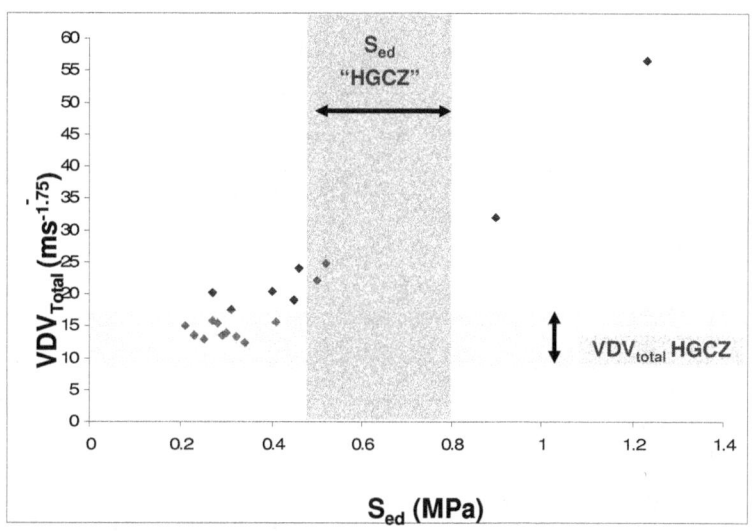

Figure 1.—Comparison of predicted health risks for 19 LHD vehicle operators based on VDV_{total} values (ISO 2631-1) and S_{ed} values (ISO 2631-5). The associated HGCZ boundaries for VDV_{total} values and HGCZ limits associated with a low and high probability of adverse health effects for S_{ed} values are shown as shaded regions.

References

Alem N [2005]. Application of the new ISO 2631-5 to health hazard assessment of repeated shocks in U.S. army vehicles. Ind Health *43*:403–412.

Cooperrider NK, Gordon JJ [2006]. Shock and impact on North American locomotives evaluated with ISO 2631 parts 1 and 5. In: Proceedings of the First American Conference on Human Vibration (June 5–7, 2006). Morgantown, WV: U.S. Department of Health and Human Services, Public Health Service, Centers for Disease Control and Prevention, National Institute for Occupational Safety and Health, DHHS (NIOSH) Publication No. 2006–140, pp. 77–78.

Eger T, Salmoni A, Cann A, Jack R [2006]. Whole-body vibration exposure experienced by mining equipment operators. Occup Ergon *6*(3/4):121–127.

ISO [1997]. Mechanical vibration and shock: evaluation of human exposure to whole-body vibration. Part 1: General requirements. Geneva, Switzerland: International Organization for Standardization. ISO 2631-1:1997.

ISO [2004]. Mechanical vibration and shock: evaluation of human exposure to whole-body vibration. Part 5: Method for evaluation of vibration containing multiple shocks. Geneva, Switzerland: International Organization for Standardization. ISO 2631-5:2004.

Johanning E, Fischer S, Christ E, Gores B, Luhrman R [2006]. Railroad locomotive whole-body vibration study: vibration, shocks, and seat ergonomics. In: Proceedings of the First American Conference on Human Vibration (June 5–7, 2006). Morgantown, WV: U.S. Department of Health and Human Services, Public Health Service, Centers for Disease Control and Prevention, National Institute for Occupational Safety and Health, DHHS (NIOSH) Publication No. 2006–140, pp. 150–151.

Index of Authors

Adewusi, S.A.	76	Johnson, C.	98
Aledo, V.	65	Johnson, P.W.	63, 115
Amirouche, F.	21, 65, 106, 122	Jurcsisn, J.G.	19
Bain, J.	10, 96	Kao, D.	94
Basista, Z.	45	Kashon, M	12
Bazrgari, B.	34, 39	Kasra, M.	34
Blood, R.	63, 115	Kaulbar, U.	68
Boileau, P. É.	57, 76, 117, 119	Khokhlov, T.	50
Boutin, J.	23, 57	Kim, H-J.	59
Bowden, D.R.	19	Kim, J.	72
Brammer, A.J.	100	Krajnak, K.	8, 12, 98, 128
Cherniack, M.G.	100	Książek, M.A.	45
Contratto, M.S.	28, 30	Larocque, R.	74
de Carvalho, L.A.P.	113	Larson, R.	124
DeMont, R.	61	Loffredo, M. A.	94
Dickey, J.	117, 119, 134	Lundstrom, R.	100
Donadon, J.V.	132	Lutz, T.	108
Dong, R.G.	70, 78, 89	Maeda, S.	15, 48
Du, J.C.	28	Manno, M.	8
Duarte, M.L.M.	113, 132	Mansfield, N.J.	54
Dudnyk, I.	126	Marcotte, P.	23, 76
Eger, T.	117, 119, 134	Martin, B.J.	59
Fischer, M.	108	Matloub, H.S.	94
Flavahan, N. A.	128	Mayton, A.G.	21, 26
Flavahan, S.	128	McDowell, T.W.	70, 78, 89
Frings-Dresen, M.H.W.	83, 130	Medeiros, E.B.	113
Girard, S.A.	74	Mendes, R.	52
Goel, V.K.	54	Miller, G. R.	12, 98, 128
Gonzalez, M.	106	Miller, R.E.	26
Govindaraju, S.	10, 92, 96	Morioka, M.	111
Grenier, S.	119, 134	Morse, T.	100
Griffin, M.J.	102, 111	Mozaffarin, A.	37
Hancock, R.	54	Neely, G.	100
Hansen, D.	124	Nélisse, H.	23, 57
Hofmann, J.	30, 41	Nikanov, A.	50
Holland, C.L.	85	Nilsson, T.	100
Holland, J.P.	85	Oddo, R.	23
House, R.	8	Oliveira, E.A.	132
Hughes, J.M.	128	Oliver, M.L.	119
Hulshof, C.T.J.	83, 130	Øvrum, A.	50
Izumi, R.	113	Pankoke, S.	30, 37
Jetzer, T.	87	Paschold, H.W.	32, 81
Jobes, C.	26	Patra, S.K.	57

Pattnaik, S.P.	72		Talykova, L.	50
Pavić, A.	104		Tarnowski, J.	45
Perottino, D.	122		Tétrault, S.	74
Peterson, D.	100		Tiemessen, I.J.H.	83, 130
Pielemeier, W.J.	43		Toppila, E.	100
Ploger, J.	63, 115		Turcot, A.	74
Pranesh, A.	61		Vaktskjold, A.	50
Rakheja, S.	57, 61, 76		Valero, B.	21, 65, 122
Riley, D. A.	10, 92, 94, 96		Vellani, J.	54
Rogers, K.	92		Warren, C.	70, 78, 89
Roy, V.	74		Warren, N.	100
Santos, M.A.R.	113		Waugh, S.	12, 98
Shibata, N.	15, 48		Webster, J.	85
Shirazi-Adl, A.	34, 39		Welcome, D.E.	70, 78, 89
Silva, L.F.	52		Wirth, O.	128
Siqueira, B.C.	113		Wölfel, H.P.	41
Skandfer, M.	50		Wu, J.	78
Smith, S.D.	19		Xu, X.	70, 89
Soma, E.	122		Yan, J.	94
Starkman, J.	106		Ye, Y.	102
Stayner, R.	17		Zahm, C.	92
Stevenson, J.	117		Zhang, L	94
Striegel, A.	30		Ziemiański, D.	45
Sutinen, P.	100		Živanović, S.	104
Syurin, S,	50			

www.ingramcontent.com/pod-product-compliance
Lightning Source LLC
Chambersburg PA
CBHW080254180526
45167CB00006B/2523